Lecture Notes in Mathematics

A collection of informal reports and seminars
Edited by A. Dold, Heidelberg and B. Eckmann, Zürich

310

Birger Iversen
University of Aarhus, Aarhus/Danmark

Generic Local Structure of the Morphisms in Commutative Algebra

Springer-Verlag
Berlin · Heidelberg · New York 1973

AMS Subject Classifications (1970): 13-02, 14-02, 14 A 05, 14 A 10

ISBN 3-540-06137-1 Springer-Verlag Berlin · Heidelberg · New York
ISBN 0-387-06137-1 Springer-Verlag New York · Heidelberg · Berlin

Offsetdruck: Julius Beltz, Hemsbach/Bergstr.

TABLE OF CONTENTS

Chapter III ETALE MORPHISMS

Chapter IV SOME FUNDAMENTAL THEOREMS

INTRODUCTION

These notes treat in the framework of commutative algebra the structure of the morphisms between algebraic varieties. The corresponding notion in differential geometry is C^{∞}-maps for which the tangent maps $T_x(f)$ has maximal rank. There are obviously three possibilities

1) $T_x(f)$ injective
2) $T_x(f)$ surjective
3) $T_x(f)$ bijective

The corresponding notions in commutative algebra are

1) unramified morphism
2) smooth morphism
3) etale morphism

These classes of morphisms are treated in Chapter I-III respectively. Chapter IV which could be read first contains three fundamental theorems in commutative algebra

Hilbert's Nullstellensatz
Zariski's Main Theorem
Chevalley's Constructibility Theorem

A certain amount of applications and ramifications of the general theory are added at the end of each chapter and I recommend the reader to study the Leitfaden to get a picture of the main body of the theory.

The material is classical in its substance but the general version we possess today owes most to Grothendieck.

Although the framework of these notes is commutative algebra I have made an effort to tie the algebra to the geometry and the analysis (Appendices to II-IV).

I have chosen to present the material in terms of commutative algebra since I have obtained essential simplifications by using more purely algebraic ideas (notably the constant application of I. 2.7 in III and the beginning of Chapter II).

"Or, au stade actuel de son dévéloppement, l'"algèbre" est souvent une transcription de résultats et malheureusement aussi de méthodes "géométriques" (J.P. Serre, 1957).

The prerequisites for reading these notes are basic commutative algebra as exposed in the work of N. Bourbaki, see Bibliography (flatness, localization, integral dependance and the theory of regular local rings for the second half part of II). The notion of separable field extension plays such a fundamental role in the theory, that I have developed it from scratch.

The main part of these notes has originated from lectures given at the University of Zurich and University of Aarhus.

I should like to thank Verner Beck and Sven Toft Jensen for a number of suggestions and for help with finishing the manuscript. I also thank Vreni Schkölziger for careful typing of the manuscript.

Zurich, January 1972

Chapter I UNRAMIFIED MORPHISMS

I.1. Derivations and differentials

Throughout these notes we place ourselves in the category of commutative rings with 1-element.

Definition 1.1 Given a morphism $f: A \to B$ and a B-module N. An A-derivation from B to N is an A-linear map $d: B \to N$ such that

$$\forall b_1, b_2 \in B \quad d(b_1 b_2) = b_1 d(b_2) + b_2 d(b_1) .$$

(d,N) is called a universal A-derivation if for any other A-derivation (d',N') there exists a unique B-linear map $v: N \to N'$ such that $d' = v \circ d$.

1.2 Construction of $(d_{B/A}, \Omega^1_{B/A})$ Let Δ $(=\Delta_{B/A})$ denote the kernel of the diagonal morphism $B \otimes_A B \to B$ $(b_1 \otimes b_2 \to b_1 b_2)$. $\Omega^1_{B/A} = \Delta/\Delta^2$ is a B-module (the two B-module structures on Δ/Δ^2 coincide).

$$d_{B/A} : B \mapsto \Omega^1_{B/A}$$

denotes the map induced by $b \to b\otimes 1 - 1\otimes b$. $d_{B/A}$ is an A-derivation in virtue of the formula

$$xy\otimes 1 - 1\otimes xy = (x\otimes 1 - 1\otimes x)1\otimes y + x\otimes 1(y\otimes 1 - 1\otimes y)$$

Proposition 1.3 Let $f: A \to B$ be a morphism, then $(d_{B/A}, \Omega^1_{B/A})$ is a universal A-derivation.

Proof: Let $d: B \to N$ be an A-derivation. We want to prove that there exists a unique B-linear map $v: \Omega^1_{B/A} \to N$ such that $d = v \circ d_{B/A}$.

Existence: Define $v'': B\otimes_A B \to N$ by $v''(a\otimes b) = bd(a)$. View $B\otimes_A B$ as B-module via second factor. This makes v'' B-linear. Let v' denote the restriction of v'' to Δ. We are going to show that the restriction of v' to Δ^2 is zero. Let $\sum x_t \otimes y_t$ $(x_t, y_t \in B)$ be an element of Δ. Then the trivial formula

$$\sum x_t \otimes y_t = \sum (x_t \otimes 1 - 1\otimes x_t) 1\otimes y_t + 1\otimes \sum x_t y_t$$

shows that the B-module Δ is generated by tensors of the form $x\otimes 1 - 1\otimes x$, thus the B-module Δ^2 is generated by tensors of the form

$$\begin{aligned}&(x\otimes 1 - 1\otimes x)(y\otimes 1 - 1\otimes y)\\ &= xy\otimes 1 - x\otimes y - y\otimes x + 1\otimes xy\end{aligned}$$

if we apply v' to this tensor we get

$$d(xy) - ydx - xdy + xyd1 ,$$

LEITFADEN

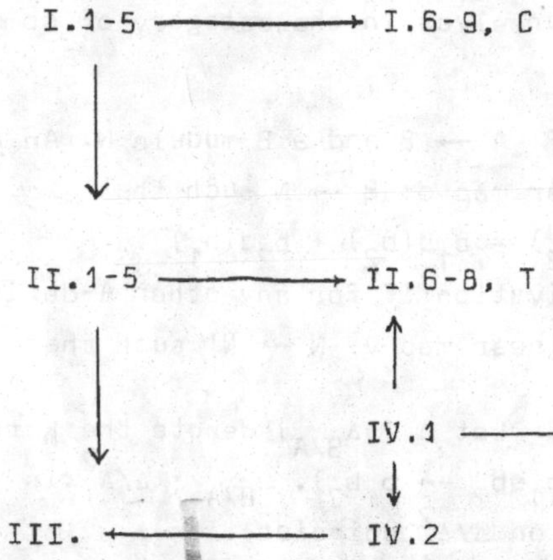

which is zero. Thus v' factors through $\Delta/\Delta^2 = \Omega^1_{B/A}$. Let v be the factorization.

$$v'(b\otimes 1-1\otimes b) = d(b)-bd(1) = d(b)$$

whence $d = v\circ d_{B/A}$.

Uniqueness: We have seen that the $B\otimes_A B$ — module Δ is generated by tensors of the form $b\otimes 1-1\otimes b$. Thus $\Omega^1_{B/A}$ is generated as B-module by the image of $d_{B/A}$. Q.E.D.

Let $k \to A$ and $f: A \to B$ be morphisms, $d_{B/A}: B \to \Omega^1_{B/A}$ is a k-derivation thus there results a B-linear map

$$\Omega^1_{B/k} \to \Omega^1_{B/A}.$$

$d_{B/k}\circ f: A \to \Omega^1_{B/k}$ is a k-derivation thus there results a B-linear map

$$\Omega^1_{A/k}\otimes_A B \to \Omega^1_{B/k}$$

which we shall denote $\underline{d_k f}$ (or $d f$, $d_k(B/A)$, $d(B/A)$ when no confusion is possible).

<u>Proposition 1.4</u> Let $k \to A$ and $f: A \to B$ be morphisms. Then the following sequence of B-modules

$$\Omega^1_{A/k}\otimes_A B \to \Omega^1_{B/k} \to \Omega^1_{B/A} \to 0$$

is exact.

<u>Proof</u>: (hint) Show that the projection of $\Omega^1_{B/k}$ onto the cokernel of $d_k f$ composed by $d_{B/k}$ is a universel A-derivation.

<u>Proposition 1.5</u> Let k be a commutative ring, I a set and $k[X] = k[X_i]_{i\in I}$. Then $\Omega^1_{k[X]/k}$ is a free $k[X]$-module with basis $(d_{k[X]/k}(X_i))_{i\in I}$.

<u>Proof</u>: (hint) Let Ω be the free $k[X]$-module on generators $(\omega_i)_{i\in I}$ and $d: k[X] \to \Omega$ the map given by $d(P) = \sum \frac{\partial P}{\partial X_i}\omega_i$. Show that (d,Ω) is a universal k-derivation.

<u>Proposition 1.6</u> Let $P_j \in k[X]$, $j\in J$, $A = k[X]/\{P_j\}_{j\in J}$.

Then
$$\Omega^1_{k[X]/k} \to \Omega^1_{A/k}$$

is surjective with kernel generated by $\{P_j dX_i\}_{i,j}$ and $\{dP_j\}_j$.

<u>Proof</u>: (hint) Let Ω^1 denote the <u>quotient</u> of $\Omega^1_{k[X]/k}$ by the submodule generated by the above elements. Use 1.5 to show that Ω^1 and the

the composit of $d_{k[X]/k}$ with the projection of $\Omega^1_{k[X]/k}$ onto Ω^1 factors through A and that the factorization is a universal k-derivation.

Proposition 1.7 Let $k \to A$ be a morphism and S a multiplicative subset of A. Then there exists a unique k-derivation $d: S^{-1}A \to \Omega^1_{A/k} \otimes_A S^{-1}A$ such that

$$d(a/s) = d_{A/k}(a)/s - ad_{A/k}(s)/s^2$$

Proof: Follows easily from the following "Let d be a derivation from A to an $S^{-1}A$-module N, then d has a unique factorization through $A \to S^{-1}A$". Proof: The main point is that the expression $d(a/s) = d(a)/s - ad(s)/s^2$ is well defined. Suppose $a/s = b/t$. Then we can find $r \in S$ such that $rta = rsb$. Apply d to this equation, multiply by st, and the result drops out. Q.E.D.

Proposition 1.8 Let $k \to A$ and $k \to k'$ be morphisms. Then, the map $A \otimes_k k' \to \Omega^1_{A/k} \otimes_k k'$ given by $a \otimes x \mapsto d_{A/k}(a) \otimes x$ is a universal k'-derivation

Proof: Left to the reader.

Proposition 1.9 Let $k \to A$ and $k \to B$ be morphisms, then the map

$$A \otimes_k B \to (\Omega^1_{A/k} \otimes_k B) \oplus (A \otimes_k \Omega^1_{B/k})$$

$a \otimes b \mapsto (d_{A/k}(a) \otimes b, a \otimes d_{B/k}(b))$

is a universal k-derivation.

Proof: Left to the reader.

Definition 1.10 Let A be a commutative ring and E an A-module. $D_A(E)$ denotes the A-algebra whose underlying A-module is $A \oplus E$, multiplication given by $(a,e)(b,f) = (ab, af+be)$. The first projection $D_A(E) \to A$ is an A-algebra homomorphism and is denoted the canonical augmentation. The kernel of the canonical augmentation has square zero and may be identified with E.

Exercise 1.11 Let $k \to A$ be a morphism and E an A-module. Show that the k-sections to $D_A(E) \to A$ may be identified with the k-derivations from A to E.

Exercise 1.12 Let $k \to A \to B$ be morphisms and I an ideal in B with $I^2 = 0$. Show that the k-liftings of $A \to B/I$ to B may be identified with the k-derivations from A to I.

Exercise 1.13 A morphism $k \to A$ is called formally unramified if for all pairs (B,I) where B is a k-algebra and I an ideal in B with $I^2 = 0$, any k-morphism $f: A \to B/I$ admits at most one k-lifting to B. Show that $k \to A$ is formally unramified if and only if $\Omega^1_{A/k} = 0$. Show moreover, that if $k \to A$ is formally unramified, if B is a k-algebra and $I \subseteq B$ a nilpotent ideal, then a k-morphism from A to B/I admits at most one k-lifting to B.

Exercise 1.14

i) The composit of two formally unramified morphisms is formally unramified.

ii) Let $f: A \to B$ and $g: B \to C$ be morphisms. $g \circ f$ formally unramified $\Longrightarrow$ g formally unramified.

iii) Let $f: A \to B$ and $A \to A'$ be morphisms, f formally unramified $\Longrightarrow$ $f \otimes 1: A' \to B \otimes_A A'$ formally unramified.

iv) Let $f: A \to B$ and $A \to A$ be morphisms. $A' \to A'$ faithfully flat and $f \otimes 1: A' \to B \otimes_A A'$ unramified $\Longrightarrow$ $f: A \to B$ formally unramified.

v) An epimorphism in the category of rings is formally unramified.

vi) Let $f: A \to B$ be a morphism. Show that f is formally unramified if and only if $f_q: A_{f^{-1}(q)} \to B_q$ is formally unramified for all prime ideals, resp. maximal ideals, q of B.

vii) The direct product of two formally unramified k-algebras, is formally unramified.

Exercise 1.15 Let k denote a field and A a local k-algebra with maximal ideal m. Suppose the composit $k \to A \to A/m$ is an isomorphism. The hereby obtained map $A \to k$ we denote $a \mapsto a(1)$. For $a \in A$ let $d(a)$ denote the residue class of $a-a(1)$ in m/m^2. Show that d is a k-derivation and that $m/m^2 \simeq \Omega^1_{A/k} \otimes_A A/m$.

I.2 Unramified morphisms

We discuss the basic properties of morphism with vanishing differentials.

Recall, that a morphism $f: A \to B$ is said to be essentially of finite type if it can be factored $A \to C \to B$ where $A \to C$ is of finite type and $C \to B$ is C-isomorphic to $C \to S^{-1}C$ for some multiplication subset S of C.

Definition 2.1 A morphism $f: A \to B$ is called unramified if $\Omega^1_{B/A} = 0$ and f is essentially of finite type.

We shall leave to the reader to verify by means of I.1, that we have the following list of properties

2.2.i. If A is a ring and S a multiplicative subset of A, then $A \to S^{-1}A$ is unramified. A surjective morphism is unramified.

2.2.ii. The composit of two unramified morphisms is unramified.

2.2.iii. If $A \to B$ is unramified and $A \to A'$ is any morphism, then $A' \to A'\otimes_A B$ is unramified.

2.2.iv. If $A \to B$ is essentially of finite type and $A \to A'$ is faithfully flat, then $A' \to A'\otimes_A B$ unramified $\Rightarrow$ $A \to B$ unramified.

2.2.v. If $A \xrightarrow{f} B \xrightarrow{g} C$ are morphisms, then gof unramified $\Rightarrow$ g unramified.

2.2.vi. Let $A \to B$ be essentially of finite type. If for all prime ideals p*) of A, $k(p) \to B\otimes_A k(p)$ is unramified, then $A \to B$ is unramified.

2.2.vii. The direct product of two unramified algebras is unramified.

Proposition 2.3 Let $f: A \to B$ be a morphism essentially of finite type. Then f is unramified if and only if the kernel of the diagonal $B\otimes_A B \to B$ is generated by an idempotent.

Proof: The kernel of $B\otimes_A B \to B$ $(b_1\otimes b_2 \mapsto b_1 b_2)$ is denoted Δ. We have $\Delta/\Delta^2 = \Omega^1_{B/A}$ (1.3). Thus if Δ is generated by an idempotent we have $\Omega^1_{B/A} = 0$.

Conversely, assume $\Delta = \Delta^2$. Δ is a finitely generated $B\otimes_A B$-module: we can find elements $x_1, \dots, x_n \in B$ such that whenever two morphisms $r,s: B \to C$ with rof = sof coincides on $x_1, \dots, x_n$ then they are equal.

We claim that $x_i\otimes 1 - 1\otimes x_i$ generates the ideal Δ.

Let Δ' denote the ideal generated by these elements and let q denote the projection of $B\otimes_A B$ onto $B\otimes_A B/\Delta'$. The two maps $b \mapsto q(b\otimes 1)$, $b \mapsto q(1\otimes b)$ have the same composit with f and coincide on $x_1, \dots, x_n$ thus they are identical. We conclude the proof with

Lemma 2.4 Let C be a commutative ring and $I \subseteq C$ a finitely generated ideal. $I^2 = I$ implies that I is generated by an idempotent.

Proof: If C is local, 2.4 follows from Nakayamas Lemma. The general case follows from the local case by Bourbaki, Alg. Comm. II, §3, n°3, cor.1 de la prop. 12. Q.E.D.

*) k(p) denotes the fraction field of A/p.

<u>Proposition 2.5</u> Let $f: A \to B$ be an unramified morphism. Let $r,s: B \to A$ be morphisms such that $r \circ f = s \circ f = 1_A$. Let finally $x: A \to k$ be a morphism from A to a <u>field</u> k and assume A is <u>connected</u> (i.e. without non trivial idempotents). Then $x \circ r = x \circ s \Rightarrow r = s$.

<u>Proof:</u> Consider the following diagram

$$\begin{array}{c} B \\ \uparrow \\ B \otimes_A B \end{array} \xrightarrow[(r,s)]{} A \xrightarrow[x]{} k$$

where $B \otimes_A B \to B$ is the diagonal morphism and apply the following remark.

Let $C \to B$ be a surjective morphism whose kernel is generated by an idempotent, $f: C \to A$ a morphism and $x: A \to k$ a morphism from A into a <u>field</u>. Assume A is connected. Then, f can be factored through $C \to B$ if and only if $x \circ f$ can be factored through $C \to B$.

<u>Proof:</u> Straight forward.

<u>Lemma 2.6</u> Let $f: A \to B$ be a morphism. Then, the kernel of the diagonal morphism $B \otimes_A B \to B$ is generated by an idempotent if and only if $\exists\, t \in B \otimes_A B$ such that

i $\quad b \otimes 1\, t = t\, 1 \otimes b$ for all $b \in B$

ii $\quad$ the image of t by the diagonal morphism is 1.

<u>Proof:</u> Obvious.

<u>Proposition 2.7</u> Let $A \to B$ be an unramified morphism. If a B-module N is flat as A-module then N is flat as B-module. - More generally, if N is a B-module, then N is (as B-module) isomorphic to a direct summand of $B \otimes_A N$, considered a B-module via first factor.

<u>Proof:</u> It suffices to prove the last statement. Let $t = \sum b_i \otimes b_i'$ ($b_i, b_i' \in B$) be as in Lemma 2.6. Then $n \mapsto \sum b_j \otimes b_j' n$ defines a map $N \to B \otimes_A N$ which is B-linear according to 2.6,i and a section to $B \otimes_A N \to N$ ($b \otimes n \mapsto bn$) according to 2.6,ii. Q.E.D.

<u>Lemma 2.8</u> Let $A \to B$ be an unramified morphism. If $(b_i)_{i \in I}$ is a basis for B as A-module, then I is finite.

<u>Proof:</u> By 2.6 we can find a tensor

$$\sum_{i \in J} x_i \otimes b_i \qquad (x_i \in B),\ J \subseteq I \text{ a finite subset, with the proper-}$$

ties i and ii of Lemma 2.6. We are going to prove that $(x_i b_j)_{i,j \in J \times J}$ generates B as an A-module. - For $b \in B$ we have

$$\sum_i bx_i \otimes b_i = \sum_i x_i \otimes b_i b$$

write $b_i b = \sum_j a_{ij} b_j$ $(a_{ij} \in A)$, then $\sum_i bx_i \otimes b_i = \sum_{i,j} x_i \otimes a_{ij} b_j = \sum_j (\sum_i a_{ij} x_i) \otimes b_j$

whence $$bx_j = \sum_i a_{ij}\, x_i$$

From $\sum_j x_j b_j = 1$ we get

$$b = \sum_j bx_j b_j = \sum_{ij} a_{ij} x_i b_j \qquad \text{Q.E.D.}$$

Exercise 2.9 Let $k \to A$ be morphism. Suppose $x_1, \ldots, x_n \in A$ generates A as k-algebra. Let $f_1, \ldots, f_n \in k[X_1, \ldots, X_n]$ and suppose

$$f_i(x_1, \ldots, x_n) = 0 \qquad \text{all } i$$

and that

$$\det_{ij} \frac{\partial f_i}{\partial X_j}(x_1, \ldots, x_n)$$

is a unit in A. Show that $k \to A$ is unramified.

Exercise 2.10 Let k denote a commutative ring and $f_1, \ldots, f_m \in k[X_1, \ldots, X_n]$. Show that

$$k[X_1, \ldots, X_n] \,/\, (f_1, \ldots, f_m)$$

is unramified if and only if for all $i = 1, \ldots, n$ the ideal generated by

$$f_1, \ldots, f_m, \frac{\partial}{\partial X_i} f_1, \ldots, \frac{\partial}{\partial X_i} f_m \text{ is } k[X_1, \ldots, X_m]$$

Exercise 2.11 Let O denote a local ring with residue field k. Let $f \to \bar{f}$ denote the canonical map $O[X] \to k[X]$. Show that if $f \in O[X]$ and $x \in k$ are such that $\bar{f}(x) = 0$ and $\bar{f}'(x) \neq 0$ then there exists at most one $a \in O$ such that $\bar{a} = x$ and $f(a) = 0$.

Exercise 2.12 Let X denote a top space, $a_o, \ldots, a_r$ continuous functions from X to $\mathbb{R}$. Define

$$f: \mathbb{R} \times X \to \mathbb{R} \text{ by}$$

$$f(t,x) = \sum_{i=o}^{n} a_i(x) t^{n-i}$$

Put $V = \{(t,x) \mid f(t,x) = 0\}$

Let $\pi: V \to X$ danote the second projection. Let s_1 and s_2 be two continuous section to π. Show that if x_o is such that $s_1(x_o) = s_2(x_o) = (t_o, x_o)$ and $\frac{\partial}{\partial t} f(t_o, x_o) \neq 0$. Then, $s_1 = s_2$ in a neighbourhood of x_o. (hint use 2.11)

I.3 Unramified algebras over a field

Let us recall a basic Lemma.

Lemma 3.1 Let $k \to k'$ be a field extension and A a finite k-algebra of finite rank n. Then

$$\# \operatorname{Hom}_k(A,k') \leq n$$

Equality holds if and only if $A \otimes_k k' = \underbrace{k' \times \ldots \times k'}_{n}$.

Proof: Passing to $A \otimes_k k'$, we may assume $k = k'$. Decompose A into a (finite) product of local rings and the Lemma follows. Q.E.D.

Proposition 3.2 Let k denote a field and A a k-algebra, essentially of finite type. Then A is unramified over k if and only if $A \otimes_k \overline{k}$ is isomorphic to a finito direct product of copies of $\overline{k}$.

Proof: We may assume $k = \overline{k}$ by 2.2,iv. The $\overline{k}$ algobra $\overline{k} \times \ldots \times \overline{k}$ (finitely many copies) is unramified by 2.2,vii. - Assume A is unramified. Then A is of finite rank over $\overline{k}$ by 2.8. Decompose A into a finite product of local $\overline{k}$-algebras $A_1 \times \ldots \times A_n$. A_i is unramified by 2.2,i and ii. Whence we may assume A local. Let m denote the maximal ideal of A. A/m is a projective A-module by 2.7, whence $m = 0$ (alternatively: $m/m^2 \simeq \Omega^1_{A/k} \otimes_A A/m$ by 1.15, whence $m = 0$ by Nakayama). Q.E.D.

Example 3.3 A field extension $k \to k'$ is unramified if and only if it is a finite separable field extension. Note, that a separable algebraic field extension which is not finite is formally unramified but not unramified. See I.5 and II.5 for a discussion of separable field extensions in general.

Lemma 3.4 Let k denote a field. A subalgebra of an unramified k-algebra is unramified.

Proof: Let B be a subalgebra of the unramified k-algebra A. $B \otimes_k \overline{k}$ is reduced (i.e. without nilpotent elements) since $A \otimes_k \overline{k}$ is reduced. Whence, by factoring $B \otimes_k \overline{k}$ in a (finite) product of local rings, $B \otimes_k \overline{k}$ is a (finite) product of copies of $\overline{k}$. Q.E.D.

Remark 3.5 Let A be an algebra of finite rank over the field k. Then there exists a largest unramified subalgebra A^{nr} of A. Namely, by 2.2,iii and ii, the tensor product is unramified and a surjective morphism is unramified by I.14,v, whence again by 2.2,iii the subalgebra generated by two unramified subalgebras is unramified. - See I.7 for a generalization of this consideration.

Lemma 3.6 Let $k \to K$ be a field extension of finite rank. If $x \in K$, then $x^q \in K^{nr}$ for some power q of the characteristic exponent p of k.

Proof: Let P be the monic polynomial with coefficients in k, irreducible in $k[X]$ and such that $P(x) = 0$. If $P'(x) = 0$, then $P' \in (P)$ and thus either $P' = 0$ or $dgP' \geq dgP$, whence $P' = 0$. If $p = 1$, this finishes the Lemma. Assume $p>1$. $P'(x) = 0 \Rightarrow P = Q(X^p)$, $Q \in k[X]$, Q irreducible.Put $x_1 = x^p$, then either $Q'(x_1) \neq 0$ or Q may be written$Q = R(X^p)$. Keep going and arrive at an irreducible polynomial S and at y which is x raised to some power of p such that $S(y) = 0, S'(y) \neq 0$, whence $y \in K^{nr}$.

Exercise 3.7 Let $k \to A$ be an unramified algebra over the field k. Let B be a k-algebra and I a nilpotent ideal in B. Show that any k-morphism $A \to B/I$ admits a unique lifting to B. (Hint. Compare 1.13 and reduce to the case where $I^2 = 0$ and $k \to A$ is a separable field extension with a single generator)

Exercise 3.8 Let k be a field and A a k-algebra of finite rank. Show that for any a A there exists a power q of p (= characteristic exponent of k) and $b \in A^{nr}$ such that $a^q - b$ is nilpotent. - Use this to prove $A^{nr} \otimes_k k' \simeq (A \otimes_k k')^{nr}$ for any field extension $k \to k'$.

Exercise 3.9 Show that any field extension $K \to K'$ (ess.) of finite type is composed of a finite number of monogenious extensions $L \to L'$ with a generator x such that either

1) x is transcendental over L.

x is algebraic and satisfies an irreducible polynomial P such that

2) $P'(x) \neq 0$ or

3) $P = X^p - a$, $a \in L$, p the characteristic exponent of L and a not a p'th power in L ($X^p - a$ irreducible if and only if a is not a p'th power in L).

Exercise 3.10 Let $A \to B$ be morphism such that B is finitely generated free as A-module. For a basis $e_1, \dots, e_n$, $\det_{ij} Tr_{B/A}(e_i e_j)$ is called the discriminant of the basis. Show

1) The discriminant of two basis differ by the square of a unit in A

2) $A \to B$ is unramified if and only if the discriminant is a unit.

(Hint. Treat first the case when A is an algebraically closed field, using that the trace of a nilpotent linear endomorphism is 0)

Exercise 3.11 Let $t_1,\dots,t_n$ be indeterminants over $\mathbb{Z}$. The elementary symmetric polynomials $s_1,\dots,s_n \in \mathbb{Z}[t_1,\dots,t_n]$ are defined by the relation (in $\mathbb{Z}[t_1,\dots,t_n,X]$)

$$(X-t_1)\ \dots\ (X-t_n) =$$
$$X^n - s_1X^{n-1} + s_2X^{n-2} + \dots + (-1)^n s_n$$

Recall that a symmetric polynomial in $t_1,\dots,t_n$ may be written (uniquely) as a polynomial in $s_1,\dots,s_n$. Let $D_n \in \mathbb{Z}[X_1,\dots,X_n]$ denote the polynomial given by

$$D_n(s_1,\dots,s_n) = \sum_{i<j}(t_i-t_j)^2$$

Given a ring k and a **monic** polynomial P with coefficients in k. Write $P = X^n - a_1X^{n-1} + a_2X^{n-2} \pm \dots$

$D_n(a_1,\dots,a_n) \in k$ is called the discriminant of P. Show that the discriminant of P is the discriminant of $k[X]/P$ with respect to the basis $1, X, X^2, \dots, X^{n-1}$. (Hint. See any old textbook on algebra)

Exercise 3.12 Let k denote a field of characteristic 2. Let p: $k \to k$ denote the function

$$x \mapsto x^2 + x$$

Show that $k/p(k)$ classifies the unramified k-algebras of rank 2. - [A non degenerate quadratic form Q over k has even dimension, 2n. The Clifford algebra $C^+(Q)$ has center which is an unramified k-algebra of rank 2. The corresponding element in $k/p(k)$ is called the Arf invariant of Q. Explicitly let $e_1,\dots,e_n,f_1,\dots,f_n$ be a symplectic basis for the bilinear form associated with Q. The Arf invariant of Q is the class (in $k/p(k)$) of $\sum_{i=1}^{n} Q(e_i)Q(f_i)$].

I.4 Cartier's equality

Let k denote a ring, Ω^1_k is short for $\Omega^1_{k/\mathbb{Z}}$.

Theorem 4.1 Let $k \to K$ be a field extension (ess.) of finite type. Then the kernel and the cokernel of the K-linear map

$$\Omega^1_k \otimes_k K \to \Omega^1_K$$

has finite rank and

$$\mathrm{rank}_K(\mathrm{Cok}) - \mathrm{rank}_K(\mathrm{Ker}) = \mathrm{trdg}_k K$$

Proof: Let us first forward some remarks which will reduce the problem to 3 special cases. - The above linear map is denoted d(K/k).

Note 1 If $k_1 \to k_2 \to K$ are field extensions, then

$$d(K/k_2) \circ d(k_2/k_1) \otimes_{k_2} K = d(K/k_1)$$

Note 2 If $f\colon V_1 \to V_2$ and $g\colon V_2 \to V_3$ are K-linear maps of finite index. Then gof has finite index and index(gof) = index(g) + index(f) (finite index: Cok and Ker have finite rank. Index = rank Cok — rank Ker)

Thus it follows (see Exercise 3.9) that we may assume K generated by a single element, t say (let $f \in k[X]$ denote the minimal polynomial for t) and that we are in one of the following three cases

1° t transcendental over k, i.e. f = 0

2° $f'(t) \neq 0$

3° $f = X^p - a$, $a \in k$ (remark $X^p - a$ is irreducible in $k[X]$ if and only if a is not a p'th power in k)

Note 3 The K-linear dual to $\Omega^1_k \otimes_k K \to \Omega^1_K$ is

$$r\colon \mathrm{Der}(K,K) \to \mathrm{Der}(k,K)$$

where Der denotes the space of $\mathbb{Z}$-derivations, r denotes the restriction map. It is clear that 4.1 is equivalent to

$$\mathrm{rank}_K \mathrm{Ker}(r) - \mathrm{rank}_K \mathrm{Cok}(r) = \mathrm{trdg}_k K.$$

Note 4 Let L denote a field, let $d\colon k \to L$ be a derivation and let $f^d \in k[X]$ denote the polynomial obtained by applying d to the coefficients of f. Then the relation

$$y = D(t)$$

is a one to one correspondance between derivations $D\colon K \to L$ which are prolongations of d and elements y in K for which

$$f^d(t) + f'(t)\, y = 0$$

Proof: Straight forward "all generators and relations are known".

If we apply Note 4 to the zero derivation, we get that the dimension of the kernel of the restriction map r of Note 3 has dimension 1, 0, 1 in case 1°, 2°, 3°. It is also immediate from Note 4 that r is surjective in case 1° and 2°.

Case 3°: Let $d\colon k \to K$ be any derivation. The obstruction to a prolongation of d to K is d(a), according to Note 4. We are going to construct a derivation d': $k \to K$ such that $d'(a) = 1$. d' will generate the cokernel of r: Let d: $k \to K$ be any derivation. $d - d(a)d'$ takes the value 0 on a whence it may be prolongated to K. The existence of a d' as above is clearly equivalent to $d_k(a) \neq 0$ where d_k: $k \to \Omega'_k$ is the universal derivation. More generally

Lemma 4.2 Let $k \to K$ be any field extension and $a \in K$. $d_{K/k}(a) \neq 0$ if and only if $a \notin K^p k$ (the subfield of K generated by k and all p'th powers).

Proof: Put $k' = K^p k$. The minimal polynomial of a with respect to k' is $X^p - a^p$. Thus the zero derivation $k' \to K$ may be prolongated to $k'(a)$ by a derivation d_0 such that $d_0(a) = 1$ by Note 4 above. Now, consider pairs (d,F) where F is a subfield of K containing $k'(a)$ and d: $F \to K$ a derivation prolongating d_0. Order these pairs by prolongation and remark that Zorn's Lemma applies. So given a pair (F,d) as above and suppose $F \neq K$ pick $r \in K - F$. The minimal equation of r relative to F is $X^p - r^p = f(X)$; we have $f^d = d(r^p) - d_0(r^p) = 0$ since $r^p \in k'$. Thus we may prolongate d by Note 4. Q.E.D.

Remark 4.3 The same proof shows that if $k \to K$ is a field extension in characteristic zero and $a \in K$, then $d_{K/k}(a) = 0$ if and only if a is algebraic over k.

I.5 Separable field extensions

Definition 5.1 Let $k \to K$ be a field extension (ess.) of finite type. Elements $x_1,\dots,x_n \in K$ are called a separating transcendence bases, if $x_1,\dots,x_n$ are algebraically independent, and $k(x_1,\dots,x_n) \to K$ is a finite separable extension.

If $\mathrm{char}(k) = p > 0$, the k-algebra F: $k \to k$ $(F(x) = x^p)$ will be denoted $k^{(p)}$.

Proposition 5.2 Let $k \to K$ be a field extension (ess.) of finite type. The following conditions are equivalent:

1) K has a separating transcendence basis
2) $\mathrm{char}(k) = 0$
   or
   $\mathrm{char}(k) = p > 0$ and the map $K \otimes_k k^{(p)} \to K^{(p)}$
   given by $x \otimes y \mapsto x^p y$ is injective (Mac Lanes criterion)
3) $\Omega^1_k \otimes_k K \to \Omega^1_K$ is injective

4) $\mathrm{rank}_K(\Omega^1_{K/k}) = \mathrm{trdg}_k K$

<u>Remark 5.3</u> Suppose char k = p>0. The composit of $K\otimes_k k^{(p)} \to K^{(p)}$ and $K \to K\otimes_k k^{(p)}$ $(x \mapsto x\otimes 1)$ is F (raising to the p'th power). Thus $K\otimes_k k^{(p)} \to K^{(p)}$ is injective if and only if $K\otimes_k k^{(p)}$ is reduced.

Remark further, that $K\otimes_k k^{(p)} \to K^{(p)}$ injective is equivalent to: "If (a_i) is a family of elements of K linearly independent over k, then (a_i^p) remain linear independent over k".

<u>Proof:</u>

3) $\Longrightarrow$ 4) in virtue of Cartier equality (4.1)

1) $\Longrightarrow$ 3) thanks to the following triviality: if $K_1 \to K_2$ and $K_2 \to K_3$ satisfies 3), then so does the composit $K_1 \to K_3$.

4) $\Longrightarrow$ 1) Let $x_1,..,x_n$ be elements of K such that $d_{K/k}(x_1),..,d_{K/k}(x_n)$ is a basis for $\Omega^1_{K/k}$. Then $k(x_1,..,x_n) \to K$ is a finite separable extension according to 3.3. Whence $\mathrm{trdg}_k k(x_1,..,x_n) = \mathrm{trdg}_k K$ and $\mathrm{trdg}_k K = n$ by 4).

1) $\Longrightarrow$ 2) We may assume char(k) = p>0. The K-linear map $K\otimes_k k^{(p)} \to K^{(p)}$ is denoted m(K/k). Note that if $k \to K$ and $K \to L$ are field extensions, then

$$m(L/K) \circ L\otimes_K(m(K/k)) = m(L/k)$$

Thus we may assume K/k is purely transcendental or K/k finite separable. In both cases K/k stays reduced after the base change $k \to k^{(p)}$ by 3.2. Conclusion by Remark 5.3.

2) $\Longrightarrow$ 4) Let $x_1,..,x_n$ be elements of K such that $d(x_1),..,d(x_n)$ form a basis for $\Omega^1_{K/k}$. That makes $k(x_1,...,x_n) \to K$ a finite (separable) extension by 3.3. It remains to show that $x_1,...,x_n$ are algebraically independent over k. This will be shown by contradiction.

So let $f\in k[X_1,..,X_n]$ $f \neq 0$ and $f(x_1,...,x_n) = 0$ and suppose no polynomial of lower degree has these properties. $f(x.) = 0$ implies

$$\sum \partial f/\partial X_i(x.)\, d \quad (x_i) = 0$$

and whence by the minimality of the degree of f that

$$\partial f/\partial X_i = 0 \quad \text{all } i$$

1) Char k = 0. This implies f constant, and thus we have arrived at a contradiction.

2) Char(k) = p>0. $\partial f/\partial X_i = 0$ implies $f = g(X_1^p,..,X_n^p)$ for some $g\in k[X_1,...,X_n]$. Write

$$g = \sum a_{i.} X^{i.} \quad (X^{i.} = \prod X_j^{i(j)})$$

and let $I \subseteq \mathbb{N}^n$ be the set of i. 's such that $a_{i.} \neq 0$. The elements

$(x.^{i.})_{i.\in I}$ are linearly independent over k (a non trivial linear combination would produce an $h\in k[X.]$ with $h(x.) = 0$, $h \neq 0$, $dg\, h \leq dg\, g < dg\, f$). Using the end of Remark 5.3 we conclude that $(x.^{pi.})_{i.\in I}$ are linearly independent contradicting

$$f(x.) = g(x.^{p}) = 0 \qquad \text{Q.E.D.}$$

<u>Definition 5.4</u> A field extension $k \to K$ (ess.) of finite type satisfying the four equivalent conditions of 5.2 is called <u>separable</u>. A field k such that all extensions $k \to K$ are separable is called <u>perfect</u>.

<u>Cor. 5.5</u> The composit of two separable field extensions is separable. A subextension of a separable extension is separable.

<u>Proof</u>: Use 5.2,3 .

<u>Cor. 5.6</u> A field k is perfect if and only if char(k) = 0 or char(k) = p>0 and $x \mapsto x^{p}$ is bijective on k.

<u>Cor. 5.7</u> Let $k \to K$ be a field extension (ess.) of finite type and separable. Then, $x_1,\ldots,x_n \in K$ is a separating transcendence basis for K if and only if $d_{K/k}(x_1),\ldots,d_{K/k}(x_n)$ is a basis for the K-vectorspace $\Omega^1_{K/k}$.

<u>Proof</u>: $\Leftarrow$ was proved under the headline 4) $\Rightarrow$ 1) in the proof of 5.2. It remains to be shown that

$$\Omega^1_{k(x.)/k} \otimes_{k(x.)} K \to \Omega^1_{K/k}$$

is an isomorphism when $x_1,\ldots,x_n$ is a separating transcendence base. The cokernel is $\Omega^1_{K/k(x.)}$ which is zero since $k(x.) \to K$ is finite separable. Injectivity follows from 5.2, 3) and the following

<u>Cor. 5.8</u> Let $k \to K \to L$ be field extensions (ess.) of finite type. Assume $k \to L$ is separable. Then the two L-linear maps

$$\Omega^1_K \otimes_K L \to \Omega^1_L$$

and

$$\Omega^1_{K/k} \otimes_K L \to \Omega^1_{L/k}$$

have isomorphic kernels and isomorphic cokernels.

<u>Proof</u>: Quite generally we have the following exact-commutative diagram:

$$\begin{array}{ccccccc}
(\Omega^1_k \otimes_k K) \otimes_K L & \to & \Omega^1_K \otimes_K L & \to & \Omega^1_{K/k} \otimes_K L & \to & 0 \\
\downarrow & & \downarrow & & \downarrow & & \\
\Omega^1_k \otimes_k L & \to & \Omega^1_L & \longrightarrow & \Omega^1_{L/k} & \to & 0
\end{array}$$

where the top row is obtained by tensoring 1.4 by L over K. - By 5.2, 3) $\Omega^1_k \otimes_k L \to \Omega^1_L$ is injective. Conclusion by the snake Lemma. Q.E.D.

Definition 5.9 Let $k \to K$ be any field extension. $k \to K$ is called separable if all subextensions (ess.) of finite type are separable.

Exercise 5.10

a) Show that a field extension $k \to K$ is separable if and only if it satisfies 2) or 3) of 5.2.

b) Generalize 5.5 and 5.8 to separable field extensions of general type.

I.6 $\otimes$ and separability

Proposition 6.1 Let k be a field, A a reduced k-algebra and $k \to L$ a separable field extension. Then $A \otimes_k L$ is reduced.

Proof: Note first that the class of separable extensions for which 6.1 is true is stable under composition and limits. Also, it suffices to prove 6.1 for A of finite type over k. Imbedding A into $\prod A/q$ where q runs through the minimal prime ideals of A, we assume that A is an integral domain. Imbed A into its fraction field to see that we are reduced to prove 6.1 in case A is a field and $k \to L$ is $k \to k(X)$ or a finite separable extension. The first case is clear. The second follows from the fact that $A \to A \otimes_k L$ is unramified and whence a finite direct product of (separable) field extensions.

Remark 6.2 6.1 is characteristic for separable extensions by Remark 5.3.

Cor. 6.3 Let k be a perfect field. Then, if A and B are reduced k-algebras, so is $A \otimes_k B$.

Proof: By the same method as in the proof of 6.1 we may assume A is a field, and thus $k \to A$ is separable. Q.E.D.

Exercise 6.4 Show that 6.3 is characteristic for perfect fields.

Exercise 6.5 Let k be a field. A k-algebra A is called geometrically reduced if $A \otimes_k \overline{k}$ is reduced. Show that

a) the tensor product of two geometrically reduced k-algebras is geometrically reduced.

b) If $k \longrightarrow A$ is geometrically reduced and $k \longrightarrow K$ is an arbitrary field extension, then $A \otimes_k K$ is geometrically reduced over K.

c) If $k \longrightarrow A$ is a field extension, then A is geometrically reduced over k if and only if $k \longrightarrow A$ is a separable extension.

d) The tensor product of a geometrically reduced and a reduced k-algebra is reduced.

I.7 π_0 of algebras over a field

Let k denote a field and A a k-algebra essentially of finite type. The subalgebras of A which are unramified over k satisfies the following two properties (see 3.5).

1° A subalgebra of an unramified subalgebra is unramified (3.4).

2° Two unramified subalgebras are both contained in a larger unramified subalgebra.

In fact

Proposition 7.1 Let k be a field and A a k-algebra ess. of finite type. Then there exists a largest unramified subalgebra of A.

Proof: In virtue of the two remarks above, it suffices to show that the ranks of the unramified subalgebras of A are bounded. Let $E \subseteq A$ be an unramified subalgebra, $\text{rank}_k E \leq$ the number of connected components of $A \otimes_k \bar{K}$. Q.E.D.

Definition 7.2 Let A denote a k-algebra ess. of finite type. The largest unramified subalgebra of A is denoted $\pi_0(A)$.

Remark 7.3 $\pi_0(A_1 \times A_2) \simeq \pi_0(A_1) \times \pi_0(A_2)$. A is connected if and only if $\pi_0(A)$ is a field.

Proof: Let E denote an unramified subalgebra of $A_1 \times A_2$. The projection, E_i of E on A_i is unramified, 2.2,i and ii, whence $E_1 \times E_2$ is unramified, 2.2,vii, and $E \subseteq E_1 \times E_2$. This proves the first part, and that if $\pi_0(A)$ is a field, then A is connected. If $\pi_0(A)$ is not a field, then it (and A) contains a non trivial idempotent. Q.E.D.

Proposition 7.4 Let A denote a k-algebra ess. of finite type and $I \subseteq A$ a nilpotent ideal in A. Then $\pi_0(A) \longrightarrow \pi_0(A/I)$ is an isomorphism.

Proof: See 3.7

Theorem 7.5 Let k denote a field, A a k-algebra ess. of finite type and $k \to k'$ a field extension. Then the canonical morphism

$$\pi_o(A) \otimes_k k' \to \pi_o(A \otimes_k k')$$

is an isomorphism.

Proof: Note that if $k' \to k''$ is a second extension and if we compose the morphism obtained by applying $\otimes_{k'} k''$ to the above morphism with

$$\pi_o(A \otimes_k k') \otimes_{k'} k'' \to \pi_o(A \otimes_k k' \otimes_{k'} k'')$$

we obtain

$$\pi_o(A) \otimes_k k'' \to \pi_o(A \otimes_k k'')$$

From this we conclude: Let $\mathcal{C}$ denote the class of morphisms between fields for which 7.5 holds for any finite type k-algebra A. Then

a) If $k \to k'$ and $k' \to k''$ belong to $\mathcal{C}$, then the composit $k \to k''$ belongs to $\mathcal{C}$.

b) If $k \to k''$ belongs to $\mathcal{C}$, then $k \to k'$ and $k' \to k''$ belongs to $\mathcal{C}$.

Note also

c) If $k \to k'$ has a filtered set of subextensions $k \to k'_t$ belonging to $\mathcal{C}$ such that k' is the union of the k'_t 's then $k \to k'$ belongs to $\mathcal{C}$

.

Thus it suffices to treat the following three cases

1° $k \to k(X)$

2° $k \to k^s$ (k^s = separable closure of k in $\bar{k}$)

3° $k^s \to \bar{k}$

Note that in case 1° the automorphism group $X \mapsto \frac{aX+b}{cX+d}$ $(ad-bc \neq 0)$ of $k(X)$ has fixed field k.

In the second case $Gal(k^s/k)$ has k as a fixed field. Thus case 1° and 2° are easily handled by means of

"Let $k \to k'$ be a field extension, G a group of k-automorphisms of k' such that k is the fixed field under G. Let V be a vector-space over k and W a k' subvector-space of $V \otimes_k k'$. W is of the form $U \otimes_k k'$ for some k-subspace U of V if and only if W is stable under the action of G on $V \otimes_k k'$ (via second factor)" Bourbaki, Algèbre Chap.8, §4, N°5, Prop.7.

Now put $k' = k^s$ or $k(X)$ and $V = A$, then $\pi_o(A \otimes_k k')$ is obviously stable under G. Thus we have a subspace U of A such that $U \otimes_k k' = \pi_o(A \otimes_k k')$.

U is necessarily a subalgebra of A and even an unramified subalgebra by 2.2,iv. Whence $\pi_o(A) \otimes_k k' \supseteq \pi_o(A \otimes_k k')$ the other inclusion is clear

(2.2,iii).

3° The extension $k^s \to \bar{k}$. The problem is to show "If A is connected $/k^s$, then $A\otimes_{k^s}\bar{k}$ is connected". Let $e\in A\otimes_{k^s}\bar{k}$ be an idempotent, then e is of the form $f\otimes 1$: write $e = \sum a_i\otimes x_i$, let q be a power of p such that $x_i^q\in k^s$ for all i, whence $e^q = f\otimes 1$ where $f = \sum a_i^q x_i^q$. But $e^q = e$. Q.E.D.

Lemma 7.6 Let $\bar{k}$ be an algebraically closed field, C a finite type $\bar{k}$-algebra. C is connected if and only if $Z = \mathrm{Hom}_{\bar{k}}(C,\bar{k})$ is connected in the Zariski topology.

Proof: Hilbert's Nullstellensatz.

Cor. 7.7 Let k be a field, A and B finite type k-algebras. Then the canonical map

$$\pi_o(A)\otimes_k\pi_o(B) \to \pi_o(A\otimes_k B)$$

is an isomorphism.

Proof: According to 7.5 we may assume $k = \bar{k}$. It suffices then to show that if A and B are connected, then $A\otimes_k B$ is connected. Put $X = \mathrm{Hom}_k(A,k)$ $Y = \mathrm{Hom}_k(B,k)$. X and Y are topological spaces via the Zarisky topology, IV.1.7. On XxY we put the topology obtained by identifying XxY with $\mathrm{Hom}_k(A\otimes_k B,k)$. Note if $x_o\in X$ then $y\mapsto (x_o,y)$ $(Y \to X\times Y)$ is a closed immersion. Let now (x_1,y_1) and (x_2,y_2) be points on XxY. Connect (x_1,y_1) to (x_1,y_2) by $\{x_1\}\times Y$ and $(x_1,y_2$ to (x_2,y_2) by $X\times\{y_2\}$. Q.E.D.

Exercise 7.8 Let k denote a field. A finite type k-algebra A is called geometrically connected if $A\otimes_k\bar{k}$ is connected.
Show that

a) A is connected if and only if $\pi_o(A)$ is a field
b) A is geometrically connected if and only if $\pi_o(A) \simeq k$
c) The tensor product of a connected k-algebra and a geometrically connected k-algebra is connected
d) A connected k-algebra with a rational point (element of $\mathrm{Hom}_k(A,k)$) is geometrically connected
e) The tensor product of two geometrically connected k-algebras is geometrically connected.

I.8 Geometrically irreducible algebras

If $k \to k'$ is a field extension, then k is said to be separably closed in k' if k itself is the only finite separable subextension of $k \to k'$.

k is said to be a <u>separably closed</u> field if k is the only finite separable extension of k.

If F is a ring, the F_{red} denotes the residue ring of F with respect to the ideal of nilpotent elements. F is called <u>irreducible</u> if F_{red} is an integral domain.

<u>Definition 8.1</u> Let k denote a field. A k-algebra F is called <u>geometrically irreducible</u> if F_{red} is an integral domain and k is separably closed in the fraction field of F_{red} (the terminology will be justified later).

<u>Proposition 8.2</u> Let $k \to k'$ be a field extension and F a geometrically irreducible k-algebra. Then $F \otimes_k k'$ is geometrically irreducible over k'.

<u>Proof</u>: Let $\mathcal{C}$ denote the class of field extensions for which 8.2 holds. It is straight forward to verify that $\mathcal{C}$ is stable under composition and that if a field extension $k \to K$ is the limit of subextensions $k \to K_t$ which are in $\mathcal{C}$, then $k \to K$ is in $\mathcal{C}$.

Next, replace F by F_{red} to see that it suffices to prove 8.2 in case F is an integral domain. Note that 8.2 is true for F if and only if it is true for the fraction field of F. Thus we may assume that F is a field.

It results that it suffices to prove 8.2 in case F is a field and $k \to K$ is a finite algebraic extension or equal to $k \to k(X)$.

Applying 7.4 and 7.5 to F we see that $(F \otimes_k K)_{red}$ is connected and that K itself is the only unramified K-subalgebra of $(F \otimes_k K)_{red}$.

In the case where $k \to K$ is finite algebraic, then $(F \otimes_k K)_{red}$ is of finite dimension over F and whence $(F \otimes_k K)_{red}$ is a finite product of field extensions of K, and 8.2 follows in this case.

In case of the extension $k \to k(X)$ we need to show that k(X) is separably closed in F(X). Note that $F \otimes_k k(X)$ is a normal domain (i.e. integrally closed in its fraction field) since it is a fraction ring of the normal ring $F[X]$. Whence an element of F(X) which is algebraic over k(X) must belong to $F \otimes_k k(X)$. But, we know that k(X) is the only unramified k(X)-subalgebra of $F \otimes_k k(X)$. Q.E.D.

<u>Cor. 8.3</u> Let k be an algebraically closed field. If A and B are k-algebras which are integral domains, then $A \otimes_k B$ is an integral domain.

<u>Proof</u>: It suffices to treat the case where B is a field, 8.3 is then a

consequence of Proposition 8.2 and Proposition 6.1.

Exercise 8.4 Let k denote a field. Show that

a) A k-algebra F is geometrically irreducible if and only if $F\otimes_k \bar{k}$ is irreducible

b) the tensor product of two geometrically irreducible algebras is geometrically irreducible

c) the tensor product of geometrically irreducible algebras and an irreducible algebra is irreducible

d) If A is geometrically irreducible over k and $k \to k'$ is a field extension, then $A\otimes_k k'$ is geometrically irreducible over k'

e) Let k be a field. The tensor product of any two irreducible k-algebras is irreducible if and only if k is separably closed

f) Show that 8.3 is characteristic for algebraically closed fields

Exercise 8.5 Discuss base extensions of integral domains.

I.9 Irreducible components and base extension

The purpose of this section is to get a detailed picture of the behaviour of irreducible component of a finite type k-algebra under base extension. The results generalize those of I.8 and apply notably to linear algebraic groups.

Definition 9.1 Let k denote a field, A a k-algebra essentially of finite type and let S denote the complement in A to the union of the minimal prime ideals in A

$$\rho_o(A) = \pi_o(S^{-1}A)$$

Remark 9.2 $S^{-1}A$ is an Artin ring whose maximal ideals correspond 1-1 to the minimal ideals of A. By 7.4 we have

$$\pi_o(S^{-1}A) = \pi_o(S^{-1}A_{red})$$

from which we conclude

9.3 $\rho_o(A)$ is a field if and only if A is irreducible.

9.4 The canonical map

$$\pi_o(A) \longrightarrow \rho_o(A)$$

is injective (use 7.3). - This map expresses the relation between connected components of A and irreducible components of A. The further investigation is based on the results of I.7, I.8 and

9.5 Let f: B → C be a flat morphism. If q is a minimal prime ideal in C, then $f^{-1}(q)$ is a minimal prime ideal in B. If f is faithfully flat, then any minimal prime ideal in B is of the form $f^{-1}(q)$, where q is a minimal prime ideal in C. - For a proof see II.7.3.

9.6 Let k → k' be a field extension. With the notation of 9.1, there is a canonical map

$$\rho_o(A)\otimes_k k' \longrightarrow \rho_o(A\otimes_k k')$$

Namely, let $S_{k'}$ denote the complement in $A_{k'} = A\otimes_k k'$ to the union of the minimal primes. It follows from 9.5, that the image of S by $A \longrightarrow A_{k'}$ is contained in $S_{k'}$, whence a map $S^{-1}A \longrightarrow S_{k'}^{-1}A_{k'}$ which give rise to a map

$$S^{-1}A\otimes_k k' \longrightarrow S_{k'}^{-1}A_{k'}$$

and finally a map

$$\rho_o(A)\otimes_k k' \longrightarrow \rho_o(A\otimes_k k')$$

Proposition 9.7 Let k denote a field, k → k' a field extension and A a k-algebra essentially of finite type. The canonical map (9.6)

$$\rho_o(A)\otimes_k k' \to \rho_o(A\otimes_k k')$$

is an isomorphism.

Proof The proof runs precisely like that of 8.2.

a) Let D denote the class of field extension for which 9.7 holds - D is clearly stable under composition. - Let us show that D is stable under filtered limits. So let $k \to K$ be the filtered union of subextensions $k \to K_t$ which all are in D. Let S, S_K, S_t be the relevant multiplicative subsets of $A, A_K = A\otimes_k K$ and $A_t = A\otimes_k K_t$ respectively. Let us first prove that

$$(*) \qquad S_K^{-1}A_K = \varinjlim_t S_t^{-1}A_t\otimes_{K_t} K$$

Note, that $S_t^{-1}A_t\otimes_{K_t} K$ may be obtained by localizing A_K at the image S_t of S_t by $A_t \to A_K$. By 9.5 we have $S_t = S_K \cap A_t$ and whence $S_K = \bigcup_t S_t$. This proves (*) above. - We have

$$\rho_o(A_K) = \pi_o(S_K^{-1}A_K) = \varinjlim \pi_o(S_t^{-1}A_t^{-1})\otimes_{K_t} K$$

and by 7.5

$$\pi_o(S_t^{-1}A_t\otimes_{K_t} K) = \pi_o(S_t^{-1}A_t)\otimes_{K_t} K$$

and whence

$$\rho_o(A_K) = \varinjlim_t \rho_o(A_t)\otimes_{K_t} K$$

By assumption $\qquad \rho_o(A_t) = \rho_o(A)\otimes_k K_t$

and whence $\qquad \rho_o(A_K) = \rho_o(A)\otimes_k K$

b) Let us prove, that for a given field extension $k \to k'$ it suffices to prove the proposition in case A is a field.

We have with the notation of 9.1 $\rho_o(A)\otimes_k k' = \rho_o(S^{-1}A)\otimes_k k'$ and

$$\rho_o(A\otimes_k k') = \rho_o(S^{-1}A\otimes_k k') \text{ by 9.5.}$$

whence we may replace A by $S^{-1}A$ and even by $(S^{-1}A)_{red}$ according to 7.4. Thus we may assume that A is a product of field extensions of k. We shall leave to the reader to check that

9.8 ρ_o commutes with direct products. This concludes the reduction to the case where A is a field.

c) By a) and b) above we may now assume that A is a field and $k \to k'$ is either a finite extension or a purely transcendental in one variable. We leave to the reader to finish proof as in 8.2. Q.E.D.

Cor. 9.9 Let A and B be k-algebras essentially of finite type. Then

$$\rho_o(A\otimes_k B) = \rho_o(A)\otimes_k\rho_o(B)$$

Proof By 9.7 we may assume $k = \bar{k}$ in which case it results from 8.3.

Cor. 9.10 Let A be a k-algebra essentially of finite type. If all components of $A\otimes_k\bar{k}$ are irreducible, then all connected components of A are irreducible.

Proof Consider the canonical myp

$$\pi_o(A) \longrightarrow \rho_o(A)$$

This is an isomorphism, since it becomes an isomorphism after tensoring with k according to 9.7, 7.5 and the assumption Q.E.D.

Appendix I.C. Multiplicative coalgebras

In this appendix k denotes a field.

The category of unramified k-algebras has a very interesting "pro-category", the category of multiplicative coalgebras. This theory applies to commutative, linear algebraic groups notably at two points, arithmetic of tori and the splitting of a commutative, linear algebraic group (in case k is perfect) into the product of a unipotent and a multiplicative group.

This material was presented by P. Russel in a course given at Harvard in the fall of 1967.

The material is organized as follows:

i) Notation and the fundamental Lemma
ii) The largest multiplicative quotient of a coalgebra
iii) Characters and diagonalizable coalgebras
iv) Classification of multiplicative coalgebras

Notation and the fundamental lemma

By a coalgebra over k we shall understand a commutative coalgebra with a counit , i.e. a triple (E,μ,ε) where E is a vector space over k, μ: $E \longrightarrow E\otimes_k E$ a k-linear map and ε: $E \longrightarrow k$ a k-linear map, these data subjected to

1) coassociativity

2) cocommutativity

3) counity

See Bourbaki, Algèbre III, §11 (1970).

By a morphism from one coalgebra to another is understood a k-linear map which commutes with μ and ε.

We shall leave to the reader to discuss base-extension of a coalgebra, define the tensor product of two coalgebras and to show that tensor product of coalgebras is the direct product in the category of coalgebras.

Let (E,μ,ε) denote a finite dimensional coalgebra. Put

$$E^* = \mathrm{Hom}_k(E,k)$$

$$\mu^* = \mathrm{Hom}_k(\mu,k), \quad \varepsilon^* = \mathrm{Hom}_k(\varepsilon,k)$$

$(E^*,\mu^*,\varepsilon^*)$ is obviously a commutative k-algebra with unit element. - This procedure gives clearly a duality between the category of coalgebras of finite rank over k and the category of commutative algebras with 1 and of finite rank over k.

<u>Lemma C.1</u> Let E denote a coalgebra over k and V a finite subset of E. Then there exists a subcoalgebra F of E of finite rank over k and containing V.

<u>Proof</u> Since the sum of two subcoalgebras is a subcoalgebra, we may assume that V consists of one element, f say. Let $(e_i)_{i\in I}$ denote a basis for E over k, and write

$$\mu(f) = \sum_i f_i \otimes e_i$$

Let F denote the subspace of E generated by $(f_i)_{i\in I}$.

By coassociativity

$$\sum \mu(f_i)\otimes e_i = \sum f_i \otimes \mu e_i$$

$$\text{write } \mu e_i = \sum x_{ij}\otimes r_j, \quad x_{ij}\in E$$

$$\sum_i \mu(f_i)\otimes e_i = \sum_{i,j} f_i \otimes x_{ij} \otimes e_j$$

and whence $\mu(f_i)\in F\otimes E$.

By cocommutativity

$\mu(f_i)$ is invariant under the symmetry $e_1\otimes e_2 \mapsto e_2\otimes e_1$ of $E\otimes E$ and thus $\mu(f_i)\in E\otimes F$.

Since $E\otimes F \cap F\otimes E = F\otimes F$ we have $\mu(F)\subseteq F\otimes F$.

By counity

$$f = (1,\varepsilon)(\mu(f)) = (1,\varepsilon)(\ f_i \otimes e_i)$$
$$= \sum f_i \varepsilon(e_i), \text{ whence } f \in F. \qquad \text{Q.E.D.}$$

The largest multiplicative quotient of a coalgebra

Definition C.2 A coalgebra E over k is called multiplicative, if for any finite dimensional subcoalgebras F of E, F^* is an unramified k-algebra.

Remark C.3 C.1 and I.3 give immediately (by duality)

3.i A subcoalgebra of a multiplicative coalgebra is multiplicative.

3.ii If $E_1 \to E_2$ is a surjective morphism of coalgebras, then E_1 multiplicative $\Rightarrow$ E_2 multiplicative.

3.iii The tensor product of two multiplicative coalgebras is multiplicative.

3.iv Let E be a coalgebra and $k \to k'$ a field extension $E \otimes_k k'$ multiplicative $\Leftrightarrow$ E multiplicative.

Proposition C.4 Let E denote a coalgebra over k. Then, there exists a pair (E_m, φ_m) where E_m is a multiplicative coalgebra and $\varphi_m : E \to E_m$ is a surjective coalgebra morphism, such that any coalgebra morphism from E to a multiplicative coalgebra may be factored through φ_m.

Proof Reduce to the case where E is of finite rank over k by means of a limit argument based on C.1, dualize and read the result in I.3.

Remark C.5 By the same method as in the proof of C.4 one gets for the functor $E \mapsto E_m$

5.i If f: $E_1 \to E_2$ is injective, then f_m is injective.

5.ii If f: $E_1 \to E_2$ is surjective then f_m is surjective.

5.iii $E \mapsto E_m$ preserves tensorproducts

6.iv $E \mapsto E_m$ commutes with base change.

Proposition C.6 Let E be a coalgebra over k, F a multiplicative coalgebra over k and

$$f: F \to E_m$$

a coalgebra morphism. Then there exists at most one morphism g: $F \to E$ such that $f = \varphi_m \circ g$.

If k is a perfect field then $\varphi_m : E \to E_m$ has a section.

Proof Reduce to E and F of finite rank over k, dualize and apply I.3.

Exercise C.7 Let $E \neq 0$ be a coalgebra over the field k. Show that

1) $E_m \neq 0$
2) $E_m \simeq k \iff$ for all finite rank subcoalgebras F of E, F^* is a local ring with residue field k.
3) Suppose $E_m \simeq k$. Show that $E \to E_m$ has a coalgebra section.
4) Suppose $E_m \simeq k$. Show that
$$\mathrm{Card}\{e \in E \mid \mu(e) = e \otimes e, \varepsilon(e) = 1\} = 1$$
5) Suppose $E_m \simeq k$. Let 1' be the element of E which satisfies $\mu(1') = 1' \otimes 1', \varepsilon(1') = 1$. Show that if $E \neq k$, then there exists $t \in E$, $t \neq 0$ such that $\mu(t) = t \otimes 1' + 1' \otimes t, \varepsilon(t) = 0$.

Characters and diagonalizable coalgebras

Definition C.8 Let (E, μ, ε) be a coalgebra over k. Put
$$\chi(E) = \{e \in E \mid \mu(e) = e \otimes e, \varepsilon(e) = 1\}$$
$E \to \chi(E)$ defines obviously a covariant functor from coalgebras to sets.

Definition C.8 Let M denote a set and let $D(M) = D_k(M)$ denote the free k-module with bases M. Define k-linear maps
$$\mu: D(M) \to D(M) \otimes_k D(M), \quad m \to m \otimes m \, (m \in M)$$
$$\varepsilon: D(M) \to k, \quad m \to 1 \, (m \in M)$$
$D(M) = (D(M), \mu, \varepsilon)$ is a coalgebra. $M \mapsto D(M)$ defines a covariant functor from sets to multiplicative coalgebras. Note that we have a natural map
$$i_M: M \to \chi(D(M))$$

Proposition C.9 For a set M, the natural map
$$i_M: M \to \chi(D(M))$$
is a bijection.

For a set M and a coalgebra E, the canonical map
$$\mathrm{Hom}_{\mathrm{coalg.}}(D(M), E) \to \mathrm{Hom}_{\mathrm{sets}}(M, \chi(E))$$
given by $f \mapsto \chi(f) \circ i_M$, is a bijection.

Proof Straight forward.

Cor. C.10 $M \mapsto D(M)$ is fully faithful.

Lemma C.11 Let k denote a field and E a k-coalgebra. Then the elements of $\chi(E)$ are linearly independent as elements of E.

Proof We may assume that E is of finite rank over k by C.1. By duality, the elements of $\chi(E)$ may be identified with k-algebra morphisms from E^* to k. These are linearly independent over k. Q.E.D.

Definition C.12 Let k denote a field. A coalgebra E over k is called diagonalizable if E is generated as k-vector space by $\chi(E)$.

Cor. C.13 A diagonalizable coalgebra is isomorphic to D(M) for some set M.

Cor. C.14 A coalgebra E is diagonalizable if and only if any finite rank subcoalgebra is dual to a finite product of copies of k.

Cor. C.15 A subcoalgebra and a quotientcoalgebra of diagonalizable coalgebra is diagonalizable.

Cor. C.16 Let k denote a field and k^s the separable closure of k. A coalgebra E is multiplicative if and only if $E\otimes_k k^s$ is diagonalizable.

Classification of multiplicative coalgebras

k denotes a field, k^s a separable closure of k and $\pi_1 = \pi_1(k)$ denotes the Galois group of k^s/k considered a topological group via the Krull topology.

By a continuous π_1-set we understand a pair (T,γ) where T is a set, γ a continuous left action of π_1 on T, T equipped with the discrete topology. By a morphism of continuous π_1-sets we understand a π_1-equivariant set map.

Definition C.17 For a coalgebra E put

$$\bar{\chi}(E) = \chi(E\otimes_k k^s)$$

where we view $\bar{\chi}(E)$ as a continuous π_1-set, the action of π_1 is induced by the action of π_1 on $E\otimes_k k^s$ via second tensor factor.

$$E \longmapsto \bar{\chi}(E)$$

is a covariant functor from coalgebras to continuous π_1-sets.

Theorem C.18 $$E \longmapsto \bar{\chi}(E)$$

is an equivalence between the category of multiplicative k-coalgebras and the category of continuous π_1-sets.

Proof Let us first construct a covariant functor

$$M \longmapsto k\langle M\rangle$$

from continuous π_1-sets to multiplicative k-coalgebras.
Let $k^s[M]$ denote the free k^s-vector space with basis M and let $\sigma\in\pi_1$ operate on $k^s[M]$ by
$$\sigma(\sum_{m\in M} x_m m) = \sum_{m\in M} \sigma(x_m)\sigma m$$
This defines a k-linear left operation of π_1 on $k^s[M]$. $k\langle M\rangle$ denotes the space of invariants under this operation considered as a k-vector space. Let us first establish, that $k\langle M\rangle\otimes_k k^s \longrightarrow k^s[M]$
is an isomorphism.

By a continuity and a direct limit argument it suffices to treat the case where M is finite. Writing M as the disjoint union of its orbits it suffices to treat the case $M = \pi_1/\theta$ where θ is an open subgroup of π_1. Let k_θ denote the elements of k^s which are invariant under θ, identify π_1/θ with $Hom_k(k_\theta,k^s)$ and define a map
$$r:\ k_\theta\otimes k^s \longrightarrow k^s[\pi_1/\theta],\ x\otimes y \longmapsto \sum_{\sigma\in\pi_1/\theta} y\sigma(x)[\sigma]$$

r is an isomorphism, since k_θ is separable over k. Let π_1 act on $k_\theta\otimes_k k^s$ via second factor and note that r is equivariant. At this point we may replace $k^s[\pi_1/\theta]$ by $k_\theta\otimes_k k^s$ and prove
$$(k_\theta\otimes_k k^s)^{\pi_1}\otimes_k k^s \longrightarrow k_\theta\otimes_k k^s$$
is an isomorphism. Taking a k-basis for k_θ it is clear by Galois theory, that
$$(k_\theta\otimes_k k^s)^{\pi_1} \simeq k_\theta \ .$$
This finishes the proof of the fact that
$$k\langle M\rangle\otimes_k k^s \longrightarrow k^s[M]$$
is an isomorphism.

Next we are goint to put a coalgebra structure on $k\langle M\rangle$ where M is a π_1-set. But first we want to make sure that
$$M \longmapsto k\langle M\rangle$$
takes cartesian products into $\otimes$-product. - Let
$$\lambda'_{M,N}:\ k^s[M]\times k^s[N] \longrightarrow k^s[M\times N]$$
denote the k^s-linear map
$$([m],[n]) \longrightarrow [m,n],m,n\in M\ .$$
Note that $\lambda'_{M,N}(\sigma x,\sigma y) = \sigma\lambda'_{M,N}(x,y), \sigma\in\pi_1$. Thus we get a map

$$\lambda_{M,N}: k<M>\otimes_k k<N> \longrightarrow k<M\times N>$$

$\lambda_{M,N}\otimes_k k'$ can be identified with $\lambda'_{M,N}$ by our previous result, thus $\lambda_{M,N}$ is an isomorphism.

Coalgebra structure on $k<M>$: Let $\mu'_M: M \longrightarrow M\times M$ be given by $m \longrightarrow (m,m)$. This induces a map

$$k<M> \longrightarrow k<M\times M> \xrightarrow{\simeq} k<M>\otimes_k k<M>$$

The composit we call μ_M. $\varepsilon : k<M> \longrightarrow k$ is induced by $[m] \longmapsto 1$ ($k^S[M] \longrightarrow k^S$). Thus we may consider $k<M>$ as a coalgebra. That $M \longmapsto k<M>$ is an inverse to $E \longrightarrow \overline{\chi}(E)$ is now straight forward. Q.E.D.

Chapter II. SMOOTH MORPHISMS

II.1 Smooth functors

Definition 1.1 Let k denote a ring. A covariant functor

$$F: \{k\text{-algebras}\} \Longrightarrow \{sets\}$$

is called formally smooth (over k) if for all pairs (k',I) where k' is a k-algebra and I an ideal in k' with $I^2 = 0$,

$$F(k') \longrightarrow F(k'/I)$$

is surjective.

Exercise 1.2 Let $S\ell_{n,k}$ denote the functor which to a k-algebra k' assigns the set of n×n matrices with coefficients in k' and determinant 1. Show that $S\ell_{n,k}$ is smooth over k.

Definition 1.3 A morphism f: A ⟶ B is called formally smooth if $Hom_A(B, -)$ is a formally smooth functor from {A-algebras} to {sets}.

Exercise 1.4 Show that

i) The composit of two formally smooth morphisms is formally smooth.

ii) If f: A ⟶ B is formally smooth, then for any A-algebra A' $f\otimes 1: A' \longrightarrow B\otimes_A A'$ is formally smooth

iii) For any ring k, k ⟶ k[X] is formally smooth.

iv) For any ring A and any multiplicative subset S of A, $A \longrightarrow S^{-1}A$ is formally smooth.

Exercise 1.5 Let A be formally smooth over k, B a k-algebra and I a nilpotent ideal in B. Show that any k-morphism f: A ⟶ B/I may be lifted to B.

Remark 1.6 If in 1.1 "surjective" is replaced by "bijective" resp. "injective" we say that F is formally etale functor resp. formally unramified functor. Similar a morphism A ⟶ B is called formally etale resp. formally unramified if $Hom_A(B, \ )$ is formally etale resp. formally unramified.

Exercise 1.7 Show that 1.4 i,ii,iv holds with smooth replaced by etale resp. unramified.

Exercise 1.8 Let k denote a ring and C a not necessarily commutative k-algebra. Define the following functor

$$Ip_C: \{k\text{-alg.}\} \Longrightarrow \text{sets}$$

by $Ip_C(k')$ = the set of idempotents in $C\otimes_k k'$. Show that Ip_C is formally etale.

If C is free of finite rank over k, show that Ip_C is of the form $Hom_k(A,-)$ where $A \simeq k[X_1,\dots,X_m]/(f_1,\dots,f_r)$.

Exercise 1.9 Let $f\colon A \to B$ be a formally unramified morphism and $g\colon B \to C$ any morphism. Show that $g\circ f$ formally smooth resp. formally etale $\Longrightarrow$ g formally smooth resp. formally etale.

Example 1.10 (The rest of I.1 could be omitted) Let k denote a commutative ring and V a finitely generated free k-module. Let Q_1 and Q_2 denote quadratic forms on V. Define the functor

$$\underline{Isom}_k(Q_1,Q_2)\colon \{k\text{-algebras}\} \Longrightarrow \{\text{sets}\}$$

whose set of k'-valued points is the set of k'-linear automorphisms u of $V\otimes_k k'$ such that $Q_1\otimes k' \circ u = Q_2\otimes k'$.

Proposition 1.11 Let Q_1 and Q_2 be quadratic forms such that the bilinear form B_1 associated with Q_1 is non-degenerated (i.e. with discriminant which is a unit in k). Then, $\underline{Isom}_k(Q_1,Q_2)$ is formally smooth.

Remark 1.12 If $k = \mathbb{Z}$, $V = \mathbb{Z}^2$, $Q_1(x,y) = Q_2(x,y) = x^2 + y^2$, then $\underline{Isom}_{\mathbb{Z}}(Q_1,Q_2)$ is not formally smooth.

Proof We have to start with a pair (k',I) where k' is a k-algebra and I an ideal in k' with $I^2 = 0$ and show that

$$\underline{Isom}_k(Q_1,Q_2)(k') \to \underline{Isom}_k(Q_1,Q_2)(k'/I)$$

is surjective. Since everything base-extends nicely we may assume $k = k'$. An element of $\underline{Isom}_k(Q_1,Q_2)(k/I)$ can be represented as a $u\in Aut_k(V)$ such that $Q_1\circ u = Q_2$ mod I. We are looking for $g\in End_k(V)$ such that

1) $g \equiv 0 \mod I$

2) $Q_1\circ(u+g) = Q_2$.

Now, $Q_1\circ(u+g) = B_1\circ(u,g)+Q_1\circ u + Q_1\circ g$. A simple calculation shows that the restriction of Q_1 to IV is zero, so condition 2) can be rewritten (assuming 1))

2) $B_1\circ(u,g) = Q_2 - Q_1\circ u$.

Let v be the adjoint to u with respect to B_1 and put $f = v\cdot g$. We are now looking for $f\in End_k(V)$ such that

i) $f \equiv 0 \mod I$

ii) $B_1(x,f(x)) = Q_2(x)-Q_1(u(x))$

all $x \in V$.

Note that $Q_3 = Q_2-Q_i\circ u$ is a quadratic form on V which is $\equiv 0 \mod I$. Let us for a moment ignore i) and look for a solution to ii).

Lemma 1.13 Let $\mathrm{Quad}_k(V)$ denote the k-module of quadratic forms on V. Then the k-linear map

$$(*) \quad \mathrm{End}_k(V) \longrightarrow \mathrm{Quad}_k(V)$$

given by

$$f \longmapsto \begin{cases} x \longmapsto \\ B_1(x,f(x)) \end{cases}$$

is surjective.

Assuming the lemma, let us finish the proof of the proposition: Note first that $\mathrm{Quad}_k(V)$ is a free k-module, hence the following commutative diagram is exact (L denotes the kernel of (*))

$$\begin{array}{ccccccccc} 0 & \longrightarrow & L & \longrightarrow & \mathrm{End}_k(V) & \longrightarrow & \mathrm{Quad}_k(V) & \longrightarrow & 0 \\ & & \downarrow & & \downarrow r & & \downarrow s & & \\ 0 & \longrightarrow & L\otimes k/I & \longrightarrow & \mathrm{End}_k(V)\otimes k/I & \longrightarrow & \mathrm{Quad}_k(V)\otimes k/I & \longrightarrow & 0 . \end{array}$$

By the snake lemma $\mathrm{Ker}(r) \longrightarrow \mathrm{Ker}(s)$ is surjective. Q.E.D.

Proof of lemma. By localization -globalization we may assume that k is a local ring. Noting that $\mathrm{Quad}_k(V)$ is a finitely generated k-module we may by Nakayamas Lemma assume that k is a field.

1°. $\mathrm{char}(k) \neq 2$. Choose an orthogonal basis for B_1 and verify by direct calculations.

2°. $\mathrm{char}(k) = 2$. Remark that B_1 is a non-degenerate alternating form, choose a symplectic basis and verify directly. Q.E.D.

Remark 1.14 The kernel for (*) is the tangent space of $\underline{\mathrm{Aut}}_k(Q_1)$ at e (II Appendix T).

Remark 1.15 Let A_1 and A_2 be non-degenerate alternating forms on V. Similar argumentation as above shows that $\underline{\mathrm{Isom}}_k(A_1,A_2)$ is smooth, the key fact being

Lemma 1.16 Let $\mathrm{Af}_k(V)$ denote the k-module of alternating forms on V. Then the map

$$\mathrm{End}_k(V) \longrightarrow \mathrm{Af}_k(V)$$

given by

$$f \longmapsto \begin{cases} (x,y) \mapsto \\ A_1(f(x),y) + A_1(x,f(y)) \end{cases}$$

is surjective. - Again the kernel of this map is the tangent space of $\underline{\mathrm{Aut}}_k(A_1)$ at e (II Appendix T).

II.2 Infinitesimal extensions

Consider a commutative diagram in the category of rings

$$\begin{array}{ccc} k & \longrightarrow & A \\ \downarrow & & \downarrow \\ k' & \longrightarrow & k'/I \end{array}$$

where I is an ideal in k' with $I^2 = 0$. From §1 we inherit the problem of deciding whether or not there exists an $f: A \rightarrow k'$ making

$$\begin{array}{ccc} k & \longrightarrow & A \\ \downarrow & \swarrow & \downarrow \\ k' & \longrightarrow & k'/I \end{array}$$

commutative. Note, that such an f corresponds to a k-algebra section to $p_1: A \times_{k'/I} k' \longrightarrow A$. Note further, that p_1 is surjective with a kernel whose square is zero.

Definition 2.1 Let A be a k-algebra. An infinitesimal extension of A is a pair E. = (E,f), where E is a k-algebra and $f: E \rightarrow A$ a surjective k-morphism whose kernel has square zero. E. is called a trivial infinitesimal extension if f has a k-algebra section. E. = (E,f) is a versal infinitesimal extension if for any other infinitesimal extension F. = (F,f) there exists a k-morphism $r: E \rightarrow F$ with $g \circ r = f$.

Example 2.2 Let $k \rightarrow A$ be a morphism and M an A-module, $D_A(N) \rightarrow A$ is a trivial infinitesimal extension (I.1.10).

Proposition 2.3 Let A be a k-algebra, E a formally smooth k-algebra and f: E $\longrightarrow$ A a surjective k-morphism. Let I denote the kernel of f. Then

$$E/I^2 \longrightarrow A$$

is a versal infinitesimal extension of A.

Proof: Straight forward.

Cor. 2.4 Any k-algebra A has a versal infinitesimal extension.

Proof: Write A as the quotient of a polynomial algebra and remark that a polynomial algebra is formally smooth.

Cor. 2.5 A k-algebra A is formally smooth if and only if one (and whence all) of its versal infinitesimal extensions is trivial.

Proof: Follows from the remarks made in the beginning of this section. An important way of computing a versal infinitesimal extension is the following:

Example 2.6 Let B and C be algebras over the ring k and let f: B $\longrightarrow$ C be a k-morphism. View $D_C(\Omega^1_{B/k} \otimes_B C)$ as a B-algebra via

$$b \longmapsto (f(b), d_{B/k}(b) \otimes 1)$$

and C a B-algebra via f. That makes the canonical map

$$D_C(\Omega^1_{B/k} \otimes_B C) \longrightarrow C$$

a B-morphism. This defines an infinitesimal extension of the B-algebra C which we will denote E.(C/B,k).

Proposition 2.7 With the notation of Example 2.6, if C is formally smooth as k-algebra, then

$$E.(C/B,k)$$

is a versal infinitesimal extension of the B-algebra C.

Proof: Straight forward.

II.3 The (truncated) cotangent complex

Let k denote a ring and A a k-algebra. To an infinitesimal extension

E. = (E,f) of A we associate a complex c.(E.) of A-modules as follows

$$c_o(E.) = \Omega^1_{E/k} \otimes_E A$$

3.1 $$c_1(E.) = \ker(f)$$

$$c_n(E.) = 0 \quad \text{otherwise.}$$

d_o: $c_1(E.) \longrightarrow c_o(E.)$ is the map $x \mapsto d(x)\otimes 1$.

Let E. = (E,f) and F. = (F,g) be two inf.ext. of A/k and r: E $\longrightarrow$ F a morphism of extensions, i.e. r: E $\longrightarrow$ F is a k-morphism with f = r∘g, then r induces in an obvious manner a map of complexes

$$c.(r): c.(E) \longrightarrow c.(F).$$

<u>Lemma 3.2</u> Let r_1 and r_2 be morphisms from the extension E to the extension F. Then c.(r_1) and c.(r_2) are homotopic.

<u>Proof:</u> $r_1 - r_2$: E $\longrightarrow$ Ker(g) is a derivation and induces the homotopy in question.

<u>Definition 3.3</u> Let A be a k-algebra and E a versal inf.ext. of A/k. The homotopy class of c.(E), which is independent of E as a consequence of previous lemma, will be denoted by $\check{T}$.(A/k). If the k-algebra structure on A is given explicitly r: k $\longrightarrow$ A we write $\check{T}$.(r) for $\check{T}$.(A/k). $\check{T}$(A/k) is called the <u>cotangent complex</u>.

<u>Theorem 3.4</u> Let $d_o: C_1 \longrightarrow C_o$ represent $\check{T}$.(A/k). Then A/k is smooth if and only if there exists an A-linear map s: $C_o \longrightarrow C_1$ with s d_o = 1.

<u>Proof:</u> Let E = (E,f) be any inf.ext. of A/k and put C. = c.(E). Define a map from

$$\{r: A \longrightarrow E \mid r \text{ is a k-algebra hom. such that } f\cdot r = 1\}$$

to

$$\{s: C_o \longrightarrow C_1 \mid s \text{ is A-linear and } s\, d_o = 1\}$$

by

$$r \mapsto \text{the derivation induced by } 1_E - r\cdot f$$

and verify that this is a bijection. Q.E.D.

<u>Remark 3.5</u> Let A be a k-algebra, D a formally smooth k-algebra and f: D $\longrightarrow$ A a surjective k-morphism with kernel I. Then $D_{/I^2} \longrightarrow A$ is a versal infinitesimal extension by 2.3. For computational purposes it is useful to remark that $\check{T}$.(A/k) may be represented

$$I_{/I^2} \longrightarrow \Omega^1_{D/k} \otimes_D A$$

The following way of computing the cotangent complex is fundamental.

Theorem 3.6 Let k be a ring and f: B ⟶ C a k-morphism. Assume C is formally smooth /k. Then $d_k f\colon \Omega^1_{B/k}\otimes_B C \to \Omega^1_{C/k}$ represents $\check{T}.(f)$.

Proof: This is a consequence of 2.7 once we have shown (with the notation of 2.6) that

$$c.(E.(C/B,k)) = df.$$

Put $E = C\oplus(\Omega^1_{B/k}\otimes_B C)$ (dual numbers) and show $(c,\omega) \mapsto df(\omega) - d_{C/k}(c)$ is a universal B-derivation from E to C-modules.

Let us now collect the basic yoga of the cotangent complex. Proofs are omitted since they are straight forward.

3.7

a) Let k ⟶ A be a morphism and S a multiplicative subset of A. Then $\check{T}.(A/k)\otimes_A S^{-1}A = \check{T}.(S^{-1}A/k)$.

b) If A ⟶ B and B ⟶ C are morphisms and A ⟶ B is formally etale, then

$$\check{T}.(C/A) = \check{T}.(C/B)$$

c) Let A ⟶ B and A ⟶ A' be morphisms. Then

$$\mathrm{Tor}_1^A(B,A') = 0 \Longrightarrow \check{T}.(B/A)\otimes_A A' = \check{T}.(B\otimes_A A'/A')$$

d) If k ⟶ A and k ⟶ B are flat morphisms. Then

$$\check{T}.(B\otimes_k A/k) = (\check{T}.(B/k)\otimes_k A)\oplus(B\otimes_k \check{T}.(A/k))$$

e) If k ⟶ A and k ⟶ B are morphisms, then

$$\check{T}.(A\times B/k) = \check{T}.(A/k)\times\check{T}(B/k)$$

II.4 The Jacobian criterion for smoothness

Definition 4.1 A morphism A ⟶ B is called smooth if A ⟶ B is formally smooth and essentially of finite presentation (see appendix II.F)

Proposition 4.2 The composit of two smooth morphisms is smooth. If f: A ⟶ B is smooth and A ⟶ A' is any morphism, then $A' \to B\otimes_A A'$ is smooth.

Proof: Left to the reader.

Proposition 4.3 If $f: A \to B$ is a morphism essentially of finite type then $\check{T}.(f)$ may be presented $N \to P$ where P is a finitely generated free B-module and N a finitely generated B-module.

Lemma 4.4 Let B be a local ring (residue field k), $d: N \to P$ a morphism of B-modules, where P is finitely generated free and N finitely generated. Then there exists a B-linear map $s: P \to N$ with $s \circ d = 1_N$ if and only if $d \otimes 1: N \otimes_B k \to P \otimes_B k$ is injective.

Proof: The factorisation of d

$$N \to \mathrm{Im}(d) \to P$$

gives $N \otimes k \to \mathrm{Im}(d) \otimes k \to P \otimes k$ which shows that

$$N \otimes k \to \mathrm{Im}(d) \otimes k \quad \text{is bijective}$$

and $\mathrm{Im}(d) \otimes k \to P \otimes k$ is injective.

The exact sequence

$$0 \to \mathrm{Im}(d) \to P \to \mathrm{Cok}(d) \to 0$$

gives (since P is free) the exact sequence

$$0 \to \mathrm{Tor}_1(\mathrm{Cok}(d),k) \to \mathrm{Im}(d) \otimes k \to P \otimes k$$

whence $\mathrm{Tor}_1(\mathrm{Cok}(d),k) = 0.$

$\mathrm{Cok}(d)$ is a B-module of finite presentation whence $\mathrm{Cok}(d)$ is free by Bourbaki, Alg. Comm. II, §3, n° 2, Cor. 2 de la Prop. 5.
This implies $\mathrm{Im}(d)$ is free and whence that the short exact sequence

$$0 \to \mathrm{Ker}(d) \to N \to \mathrm{Im}(d) \to 0$$

is split exact, and whence that

$$0 \to \mathrm{Ker}(d) \otimes k \to N \otimes k \to \mathrm{Im}(d) \otimes k \to 0$$

is exact. Recall that $N \otimes k \to \mathrm{Im}(d) \otimes k$ is bijective, whence $\mathrm{Ker}(d) = 0$ by Nakayamas Lemma. Q.E.D.

Cor. 4.5 (Jacobian Criterion) Let $f: A \to B$ be a local morphism essentially of finite presentation and let k denote the residue field of B. f is smooth if and only if $\check{T}.(f) \otimes_B k$ is injective.

Cor. 4.6 Let k denote a field $f_1,\dots,f_r \in k[X_1,\dots,X_n]$ and $x \in k^n$ a common zero for $f_1,\dots,f_r$. Put $A = k[X_1,\dots,X_n]/(f_1,\dots,f_r)$ and let m denote the ideal $(X_1-x_1, X_2-x_2,\dots,X_n-x_n)$. Then A_m is smooth over k if and only if the rank of the matrix

$$\left\{\frac{\partial f_i}{\partial X_j}\right\}$$

is equal to the minimal number of generators for the ideal generated by $f_1,\dots,f_r$ in $k[X_1,\dots,X_n]_m$.

In particular if the rank of the above matrix is r, then A_m is smooth over k.

Proof: Follows from 4.5, Nakayamas Lemma and the following

Remark 4.7 Let C denote a local ring with residue field k and I an ideal contained in the maximal ideal of C. Then

$$I\otimes_C k \longrightarrow (I/I^2)\otimes_C k$$

is an isomorphism ($I^2\otimes k \longrightarrow I\otimes k$ is the zero map).

Proposition 4.8 Let $A \longrightarrow B$ be a morphism essentially of finite presentation.

1) If q is a prime ideal of B such that $A \longrightarrow B_q$ is smooth then there exist $r\in B-q$ such that $A \longrightarrow B_r$ is smooth.
2) If $A \longrightarrow B_m$ is smooth for all maximal ideals m of B then $A \longrightarrow B$ is smooth.

Proof: By means of 3.4 and 4.3 one can translate the problems into problems about modules. For 2) see Bourbaki, Alg. Comm. II, §3, n^o3, Cor. 1 de la prop. 12. For 1) loc. cit. §5, n^o1 .

Exercise 4.9 Let $A \longrightarrow B$ be essentially of finite presentation, $A \longrightarrow A'$ a faithfully flat morphism. Show, that $A' \longrightarrow B\otimes_A A'$ smooth $\Longrightarrow$ $A \longrightarrow B$ smooth (hint: Use 4.3 and the fact that a flat module of finite presentation is projective, Bourbaki, Alg. Comm., II, §5, n° 2, Cor. 2 de Th. 1. or the following exercise).

Exercise 4.10 Let A denote a commutative ring, E an A-module. $E^* = \mathrm{Hom}_A(E,A)$.

i) Show that E is finitely generated if and only if the canonical map $E\otimes_A E^* \longrightarrow \mathrm{End}_A(E)$ is an isomorphism (hint: If isomorphism, then there exist $x_1,\dots,x_r\in E$ and $x_1',\dots,x_r'\in E^*$ such that $\sum x_i'(e)x_i = e$ for all $e\in E$).
ii) Use i) to show that if E is flat of finite presentation then E is projective.

II.5 Cohen's Theorem

is this

Theorem 5.1 A separable field extension is formally smooth.

Before going to proof lets note the corollaries.

Cor. 5.2 Let $k \to K$ be any field extension. Then $\check{T}.(K/k)$ may be represented

$$\Omega^1_k \otimes_k K \to \Omega^1_K$$

Proof: Remark, that if L is a field and L_o its prime field, then

$$\Omega^1_L = \Omega^1_{L/L_o}.$$

Cor. 5.2 is a consequence of 3.6 and the fact that K is separable over its prime field (I.5.6).

Cor. 5.3 A field extension is formally smooth if and only if it is separable.

Proof: (of Cohen's theorem) Let us first treat the case where K/k is (ess.) of finite type. By I.5.2 it suffices to treat the case $K = k(X)$, which we leave to the reader, and the case $K = k[X]/f$ where f is irreducible in $k[X]$ and $df/dX \neq 0$. The last case we shall deal with directly. So given a pair (E,I) where E is a k-algebra and I an ideal in E with square zero and given a k-morphism $k[X]/_f \to E/I$ $(X \mapsto \bar{x})$. Pick an $x \in E$ which lifts $\bar{x}$. We are looking for $e \in I$ such that

$$f(x+e) = 0.$$

We have $$f(x+e) = f(e) + ef'(x)$$

whence we can solve for e since $f'(x)$ is a unit in E/I and then also in E.

General case: Consider an infinitesimal extension $E. = (E,f)$ of the k-algebra K. Let V denote the kernel of f, viewed as a K-vector space. Choose a k-linear map $s\colon K \to E$ such that $f \circ s = 1$. $(x,v) \to s(x) + v$ identifies $K \oplus V$ with E as k-vector space. The transport of the multiplicative structure on E to $K \oplus V$ must have the form

5.4 $$(x,v)(y,w) = (xy, b(x,y) + xw + yv)$$

where $b\colon K \times K \to V$ is a k-bilinear map which satisfies

5.5 $$b(x,y) = b(y,x) \quad \text{(commutativity)}.$$

$$b(xy,z) + zb(x,y) = xb(y,z) + b(x,yz) \text{(associativity)}$$

Conversely, a bilinear map b as above satisfying 5.5 is easily seen to define (via 5.4) an infinitesimal extension of K/k, which we shall denote E.(b). - The reader will easily work out that E.(b) is a trivial extension if and only if there exists a k-linear map c: K $\longrightarrow$ V such that

5.6 $$b(x,y) = xc(y) - c(xy) + yc(x)$$

To handle the information we build a little complex of K-vectorspaces P. = P.(K/k) as follows

$$P_i = 0 \quad \text{for} \quad i \neq 1, 2, 3$$

$$P_1 = K\otimes_k K, \qquad P_2 = K\otimes_k K\otimes_k K$$

$$P_3 = K\otimes_k K\otimes_k K\otimes_k K \oplus K\otimes_k K\otimes_k K$$

and

$$d_1(x_1\otimes x_2\otimes x_3) = x_1x_2\otimes x_3 - x_1\otimes x_2x_3 + x_3x_1\otimes x_2 + x_3x_1\otimes x_2$$

$$d_2((x_1\otimes x_2\otimes x_3\otimes x_4,\ y_1\otimes y_2\otimes x_3)) =$$

$$= x_1x_2\otimes x_3\otimes x_4 - x_1\otimes x_2x_3\otimes x_4 + x_1\otimes x_2\otimes x_3x_4 - x_4x_1\otimes x_2\otimes x_3 +$$

$$+ y_1\otimes y_2\otimes y_3 - y_1\otimes y_3\otimes y_2$$

For a given K-vectorspace V, the space of k-linear maps b: $K\otimes_k K \longrightarrow V$ which satisfies 5.5 modulo the functions of the form 5.6 is naturally isomorphic to

$$H^2(\mathrm{Hom}_K(P.,V))$$

which again is isomorphic to

$$\mathrm{Hom}_K(H_2(P.),V)$$

Consequently, K/k is formally smooth if and only if $H_2(P.) = 0$.

Now suppose k $\longrightarrow$ K separable. Write $K = \varinjlim K'$ where K' runs through all subextensions of K of (ess.) finite type. Then

$$H_2(P.(K/k)) = \varinjlim H_2(P.(K'/k))$$ Q.E.D.

II.6 Localities over a field

Definition 6.1 Let k denote a field. By a locality over k is understood a k-algebra O, essentially of finite type, such that O is a local ring.

Let us first recall the basic fact about

regular local rings 6.2. A local ring O (maximal ideal m, residue field K) is called a regular local ring if $\mathrm{rank}_K(m/m^2) = \text{Krull dim}(O)$.

a) If q is a prime ideal in a regular local ring O, then O_q is regular.

b) A regular local ring is a unique factorization domain.

c) Let $p \subset q$ be prime ideals in the regular local ring O. All saturated sequences of prime ideals $p = p_0 \subset p_1 \ldots \subset p_r = q$ have the same length.

d) For any finitely generated module M over a regular local ring O, there exists an exact sequence

$$o \rightarrow L_n \rightarrow L_{n-1} \rightarrow \ldots L_1 \rightarrow L_o \rightarrow M \rightarrow o$$

of A-modules, where L_i is finitely generated and free.

e) Let O be a regular local ring and I an ideal of O. O/I is a regular local ring if and only if there exists generators $x_1, \ldots, x_n$ (n = Krull dim O) for m such that $x_1, \ldots, x_s$ generates I.

f) Let k be a field, p a prime ideal of $k[X_1, \ldots, X_n]$. Then $k[X_1, \ldots, X_n]_p$ is a regular local ring.

<u>Theorem 6.3</u> Let k be a field, O a locality over k with residue field K.
Then i) O is smooth over k $\Rightarrow$ O is a regular local ring.
ii) Suppose K/k is a separable extension. Then O is smooth over k $\Leftrightarrow$ O is a regular local ring.

<u>Proof</u>: Choose a locality A over k of the form $k[X_1, \ldots, X_r]_p$, where p is a prime ideal in $k[X_1, \ldots, X_r]$, and a surjective local morphism $f: A \rightarrow O$. Let n denote the maximal ideal of A and I the kernel of f. We have the following commutative diagram of K-vectorspaces

$$\begin{array}{ccc} I/I^2 \otimes_A K & \xrightarrow{(1)} & \Omega^1_{A/k} \otimes_A K \\ (3) \big\downarrow & \nearrow & (2) \\ n/n^2 & & \end{array}$$

where (1) and (2) are induced by $d_{A/k}$ and (3) by the inclusion of I in n. - The theorem will result from the following observations

(1) is injective $\Leftrightarrow$ O is smooth over k. This is in fact the Jacobian Criterion 4.5

(2) is injective $\Leftrightarrow$ K is separable over k.
This is clear by 5.3 and the fact that (2) represents $\check{T}.(K/k)$

(3) is injective $\Leftrightarrow$ O is regular. Note that $I/I^2 \otimes_A K \simeq I \otimes_A K$ by Remark 4.7. Conclusion by 6.2e and Nakayamas Lemma. Q.E.D.

<u>Proposition 6.4</u> Let O be a smooth locality over the field k with maximal ideal m and residue field K. Then, K/k is separable if and only if

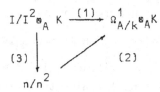

$$m/m^2 \to \Omega^1_{O/k} \otimes_O K$$

is injective.

Suppose K/k separable. Then, a basis for $\Omega^1_{O/k}$ is obtained as follows: Pick a (regular) parameter system $x_1, \ldots, x_n$ for O and elements $t_1, \ldots, t_n$ in O whose residue classes form a separating transcendence basis for K/k. Then

$$d(x_1), \ldots, d(x_n), d(t_1), \ldots, d(t_n)$$

is a basis for $\Omega^1_{O/k}$.

<u>Proof:</u>

$$m/m^2 \to \Omega^1_{O/k} \otimes_O K$$

represents $\check{T}.(K/k)$ whence the first part follows from 5.3 and 3.4. - The second part follows from the exact sequence

$$0 \to m/m^2 \to \Omega^1_{O/k} \otimes_O K \to \Omega^1_{K/k} \to 0$$

in combination with Nakayamas Lemma and I.5.7.

<u>Exercise 6.5</u> Let A be a locality over the field k (maximal ideal m, residue field K). Show, that if K/k separable, then

$$m/m^2 \to \Omega^1_{A/k} \otimes_A K \quad \text{is injective}$$

<u>Exercise 6.6</u> Let A be a locality over the field k and let K denote the residue field of A. Define

$$\text{cid}(A/k) = \text{Krull dim}A + \text{trdg}_k K - \text{index}_K \check{T}.(A/k) \otimes_A K$$

Show that if $A = O/I$ where O is a smooth locality over k, then $\text{cid}(A/k)$ = minimal number of generators for the ideal I-$\text{ht}_O(I)$. In particular

$$\text{index}_K \check{T}.(A/k) \otimes_A K \leq \text{Krull dim } A + \text{trdg}_k K$$

and equality if and only if A is a local complete intersection. (Hint: Use the commutative diagram from the proof of 6.3)

<u>Exercise 6.7</u> Let k denote a non perfect field, and let a be an element in k which is not a p'th power. Show that 1) $k[X,Y]/(X^p - aY + a)$ is geometrically irreducible.

2) The residue class $\bar{Y}$ of Y in the above ring generates a maximal ideal

3) $k[X,Y]/(X^p - aY + a)$ localized at $(\bar{Y})$ is a regular local ring

4) that the local ring in 3) is <u>not</u> smooth over k.

II.7 Dimension formulas

Theorem 7.1 Let k denote a field. O a locality over k, which is an integral domain. Let k', resp. K denote the residue field, resp. fraction field. Then

$$\text{Krull dim } O = \text{trdg}_k K - \text{trdg}_k k'$$

Proof: Choose a k-isomorphism $O \simeq A/p$ where A is a smooth locality over k. $\check{T}.(K/k)$ may be represented

$$p/p^2 \otimes_A K \to \Omega^1_{A/k} \otimes_A K$$

$\check{T}.(k'/k)$ may be represented

$$m/m^2 \to \Omega^1_{A/k} \otimes_A k' \qquad (m = \text{maximal ideal in } A)$$

Applying Cartier equality (I.4.1) two times we get

$$\text{trdg}_k K = \text{rank}_A \Omega^1_A - \text{rank}_K p/p^2 \otimes_O K$$

$$\text{trdg}_k k' = \text{rank}_A \Omega^1_A - \text{rank}_{k'} (m/m^2)$$

Let n denote the maximal ideal of A_p, note that $p/p^2 \otimes_O K \simeq n/n^2$. Thus, by 6.2 and 6.3.i,

$$\text{rank}_K \; p/p^2 \otimes_O K = \text{Krull dim } A_p$$

$$\text{rank}_{k'} \; m/m^2 = \text{Krull dim } A$$

Subtracting the two formulas above we get the Dimension formula.

Q.E.D.

Proposition 7.2 k denotes a field, A a k-algebra essentially of finite type and $k \to k'$ a field extension. Let p' be a prime ideal in $A' = A \otimes_k k'$ and let p denote the restriction of p' to A. Then

$$\text{Krull dim } A_p + \text{trdg}_k \, ff(A/p) =$$

$$\text{Krull dim } A'_{p'} + \text{trdg}_{k'} ff(A'/p')$$

Let us grant the following

Lemma 7.3 Let $f: A \to B$ be a flat morphism. Given a prime q' of B and a prime p of A contained in $f^{-1}(q')$. Then there exists a prime p' of B contained in q' such that $f^{-1}(p') = p$ (in particular the inverse image of a minimal prime is a minimal prime).

Proof of 7.2

a) Case where A = F is a field and p' a minimal prime ideal in $F' = F \otimes_k k'$. Let E be a purely transcendental subextension F such that F is finite over E. $E \otimes_k k' = E'$ is an integral domain and F' is finite and flat over E'. Whence the contraction of p' to E' is the zero ideal by 7.3 and $\mathrm{trdg}_{k'}\mathrm{ff}(F'/p') = \mathrm{trdg}_{k'}\mathrm{ff}(E') = \mathrm{trdg}_k E$.

b) Case where A is arbitrary and p' minimal.
By 7.3 p is minimal. Put F = fraction field of A/p. and p' may be identified with a minimal ideal in $F\otimes_k k'$ and the conclusion follows from a).

c) General case.

Choose a minimal prime ideal q' of A' contained in p'. Then the restriction, q of q' to A is minimal in A and contained in p. Whence by the proceeding results

$$\mathrm{trdg}_{k'}\mathrm{ff}(A'/q') = \mathrm{trdg}_k \mathrm{ff}(A/q)$$

Use 7.1 on the local rings $(A/q)_p$ and $(A'/q')_{p'}$ and 7.2 follows in general. Q.E.D.

Proof of 7.3 Put $q = f^{-1}(q')$. $A_q \longrightarrow B_{q'}$ is a flat, local morphism and whence faithfully flat by Bourbaki, Alg. Comm. I, §3, n°5, Prop. 9. This implies that

$$\mathrm{Spec}(B_{q'}) \longrightarrow \mathrm{Spec}(A_q)$$

is surjective by loc. cit. II, §2, n°5, cor 4 de prop. 11. Q.E.D.

Exercise 7.4 Let A be a locality over the field k, and let K denote the residue field of A. Show that

i) Krull dim A + $\mathrm{trdg}_k K \leq \mathrm{rank}_K \Omega^1_{A/k}\otimes_A K$

ii) The above inequality is an equality if and only if A is smooth over k.

(Hint: Assume first that K = k. Next consider the K locality A' obtained by localizing $A\otimes_k K$ at the kernel of the morphism given by $a\otimes x \longrightarrow \bar{a}x$ ($a \longrightarrow \bar{a}$ denotes the canonical projection of A onto K)).

Exercice 7.5 Let A denote a k-algebra of finite type. dimA = Krull dimA. Show (hint. Use Hilbert's Nullstellensatz IV.1 now and then)

i) $\dim A = \sup_q \mathrm{trdg}_k \mathrm{ff}(A/q)$ as q runs through all minimal prime ideals of A

ii) $\dim A = \sup_m$ Krull $\dim A_m$ as m runs through all maximal ideals of A.

iii) if $k \longrightarrow L'$ is a field extension, then

$$\dim A = \dim A\otimes_k L'$$

iv) if B is a second k-algebra of finite type, then

$$\dim A\otimes_k B = \dim A + \dim B.$$

II.8 Special criterions for smooth localities

The affine coordinate ring A for a linear algebraic group over a field k has the property that $\Omega^1_{A/k}$ is a free A-module (see II.T.11). We are going to make this assumption on a k-algebra and study the consequences (8.3 below is a generalization of Cartier's theorem: "algebraic groups in characteristic zero are smooth").

Proposition 8.1 Let A be locality over a field k. Then A is smooth over $k \Longleftrightarrow \Omega^1_{A/k}$ is free and A is geometrically reduced (i.e. $A\otimes_k \overline{k}$ is reduced).

Proof: By 4.9 we may assume $k = \overline{k}$. Thus k is perfect and we can apply the following lemma (which has a further application, see 8.3)

Lemma 8.2 Let k denote a perfect field, and A a locality over k. Assume $\Omega^1_{A/k}$ is free and that A_q is a field for all minimal primes q of A, then, A is smooth over k.

Proof: Let m denote the maximal ideal of A, K the residue field. $k \to K$ is separable since k is perfect, whence $\check{T}.(K/k)$ may be represented

$$m/m^2 \to \Omega^1_{A/k}\otimes_A K$$

whence by Cartier's equality

$$\operatorname{trdg}_k K = \operatorname{rank}_A \Omega^1_{A/k} - \operatorname{rank}_K (m/m^2).$$

For a minimal prime q of A we have

$$\Omega^1_{A_q/k} = \Omega^1_{A/k}\otimes_A A_q$$

whence

$$\operatorname{trdg}_k A_q = \operatorname{rank}_A \Omega^1_{A/k}$$

for all minimal primes q of A. Using the dimension formula 7.1 on A/q we get

$$\text{Krull dim } A/q = \operatorname{rank}_A \Omega^1_{A/k} - \operatorname{trdg}_k K .$$

Comparing this with the formula above, we get

$$\text{Krull dim } A/q = \operatorname{rank}_K m/m^2$$

for all minimal prime ideals q of A, whence by 6.2 A is regular and also smooth by 6.3.ii. Q.E.D.

Theorem 8.3 Let A be a locality over a field k of characteristic zero. Then, A is smooth if and only if $\Omega^1_{A/k}$ is a free A-module.

Proof: One way is clear (smooth algebras have always finitely generated, projective differentials). Assume $\Omega^1_{A/k}$ is free. According to the preceeding Lemma 8.2 it suffices to prove that A_q is a field whenever q is a minimal prime ideal in A. Replacing A by A_q it suffices thus to prove "If A is a locality over k of Krull dim 0, then $\Omega^1_{A/k}$ free $\Longrightarrow$ A is a field".

Proof: Let m denote the maximal ideal of A and K the residue field. Let us first show that

$$x \in m \Longrightarrow d_{A/k}(x) \in m\Omega^1_{A/k}.$$

Suppose $x^n = 0$ but $x^{n-1} \neq 0$ $(n \geq 2)$. Choose a basis $\omega_1, \ldots, \omega_r$ for $\Omega^1_{A/k}$ and write

$$d_{A/k}(x) = a_1\omega_1 + \ldots + a_r\omega_r \quad (a_i \in A).$$

$$d(x^n) = nx^{n-1}d(x) = 0$$

whence $x^{n-1}a_i = 0$ all i and whence $a_i \in m$.

Next, choose k-algebra section s to the projection $A \rightarrow A/m$. This can be done (1.5) since m is nilpotent and $k \rightarrow K$ is smooth. Now, view A as a K-algebra via s. It suffices now to show $\Omega^1_{A/K} = 0$, since this will make A unramified over K. We have an exact sequence

$$\Omega^1_{K/k} \otimes_K A \rightarrow \Omega^1_{A/k} \rightarrow \Omega^1_{A/K} \rightarrow 0$$

whence by Nakayamas lemma it suffices to prove that

$$\Omega^1_{K/k} \rightarrow \Omega^1_{A/k} \otimes_A K$$

is surjective. For $a \in A$ choose $x \in K$ such that $a - s(x) \in m$, which gives

$$d_{A/k}(a) - d_{A/k}(s(x)) \in m\Omega^1_{A/k}$$

by the above remark. Q.E.D.

II. Appendix F, relative finiteness conditions

Definition F.1 A morphism $f: A \rightarrow B$ is of finite presentation if B is A-isomorphic to an A-algebra of the form

$$A[X_1, \ldots, X_n]/(f_1, \ldots, f_r)$$

F.2 The composit of two morphisms of finite presentation is of finite presentation

F.3 If $A \rightarrow B$ is of finite presentation and $A \rightarrow A'$ is any morphism, then $A' \rightarrow B \otimes_A A'$ is of finite presentation

F.4 If $A \xrightarrow{f} B \xrightarrow{g} C$ are morphisms with $g \circ f$ of finite presentation and f of finite type, then g is of finite presentation

F.5 If $f: A \rightarrow B$ is surjective and of finite presentation, then the kernel of f is finitely generated.

Proof: F.2 and F.3 are completely trivial. F.4:
Consider the commutative diagram

$$\begin{array}{ccccc} & & b_1 \otimes b_2 & \longmapsto & b_1 b_2 \\ & & B \otimes_A B & \longrightarrow & B \\ & & \downarrow 1\otimes g & & \downarrow g \\ B & \longrightarrow & B \otimes_A C & \xrightarrow{(g,1)} & C \\ b & \longmapsto & b \otimes 1 & & \end{array}$$

The square is cocartesian (i.e. identifies C with the tensorproduct of B with $B\otimes_A C$ over $B\otimes_A B$). The diagonal map is of finite presentation (see the proof of I.2.3). Hence (g,1) is of finite presentation by F.3, also $B \to B\otimes_A C$ is of finite presentation, F.3. g is the composit of the two maps in the bottom row above, whence g is of finite presentation by F.2.

Proof of F.5: We can find a surjective A-morphism g: $A[x_1,\ldots,x_n] \to B$ whose kernel is finitely generated. We can find a morphism h: $A[x_1,\ldots,x_n] \to A$ such that the following diagram is commutative

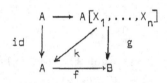

$$\begin{array}{ccc} A & \longrightarrow & A[X_1,\ldots,X_n] \\ {\scriptstyle id}\downarrow & \swarrow {\scriptstyle k} & \downarrow {\scriptstyle g} \\ A & \xrightarrow{f} & B \end{array}$$

the image by h of the kernel of g is the kernel of f. Q.E.D.

Definition F.6 A morphism f: $A \to B$ is essentially of finite present- if f can be factored $A \to C \to B$ where $A \to C$ is of finite presentation and $C \to B$ is a fraction morphism (i.e. B is C-isomorphic to $S^{-1}C$ for some multiplicative subset S of C)

F.7 The composit of two morphisms essentially of finite presentation is essentially of finite presentation

F.8 If $A \to B$ is essentially of finite presentation, then $A' \to A'\otimes_A B$ is essentially of finite presentation

F.9 If $A \xrightarrow{f} B \xrightarrow{g} C$ are morphisms with gof essentially of finite presentation, and f essentially of finite type, then g is ess. of finite presentation.

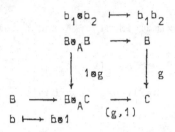

Proof: For F.7 and F.8 remark that if r: $C \to D$ is a fraction morphism and s: $D \to E$ is of finite presentation, then we can find a commutative

square

$$
\begin{array}{ccc}
C & \xrightarrow{\ r\ } & D \\
\downarrow u & & \downarrow s \\
F & \xrightarrow{\ v\ } & E
\end{array}
$$

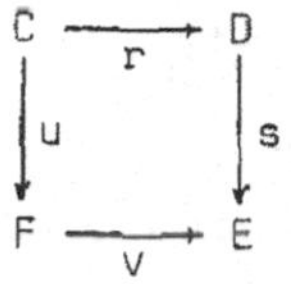

where v is a fraction morphism and u is of finite presentation. F.9:
The proof of F.9 is quite like that of F.4. One just has to remark that

F.10 If A ⟶ B is essentially of finite type, then the kernel of the
diagonal map $B\otimes_A B \longrightarrow B$ is finitely generated (see proof of I.2.3).

Recall that a morphism f: A ⟶ B is called a <u>local morphism</u> if A and B
are local rings and the inverse image of the maximal ideal in B is the
maximal ideal in A.

F.11 Let f: A ⟶ B be a surjective local morphism essentially of finite
presentation. Then the kernel of f is finitely generated.

<u>Proof of F.11</u>: We can find a prime ideal q in $A[X_1,\ldots,X_n]$ which
contracts to the maximal ideal in A and a surjective local A-morphism
g: $A[X_1,\ldots,X_n]_q \longrightarrow$ B with finitely generated kernel. Lift g to a local
morphism

$$h: A[X_1,\ldots,X_n]_q \longrightarrow A$$

such that the following diagram is commutative

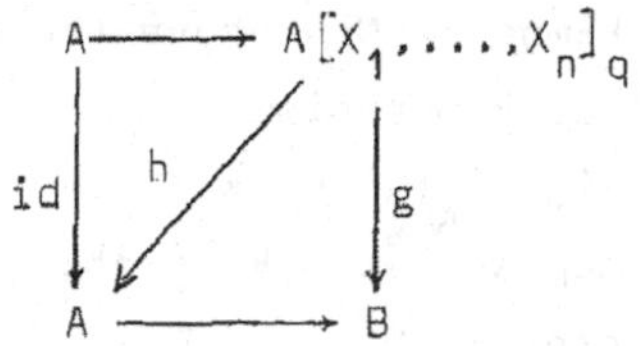

The image by h of the kernel of g is the kernel of f.   Q.E.D.

<u>Exercise</u>. Show that

F.12 Let f: A ⟶ B be a finite morphism (i.e. a morphism which makes B
a finitely generated Amodule). Then f is a morphism of finite
presentation if and only if B is an A-module of finite presentation
(i.e. exists exact sequence of A-modules

$$A^n \longrightarrow A^m \longrightarrow B \longrightarrow 0)$$

(Hint: Write B as a quotient of the tensor product of a finite number
of A-algebras of the form $A[X]/(F)$ where F is a monic polynomial).

<u>Exercise</u> (Mapping germs) Let k denote a ring, A and B two k-algebras

of finite type, q a prime ideal in B and p a prime ideal in A.

For a k-morphism $h: A \to B$ with $h^{-1}(q) = p$, $h_q: A_p \to B_q$ denotes the induced map. Show that

F.13 If f and g are k-morphisms from A to B with $f^{-1}(q) = g^{-1}(q) = p$ and $f_q = g_q$ then there exist $t \in B-q$ such that f and g induces the same map $A \to B_t$.

F.14 If $r: A_p \to B_q$ is a local morphism and A is of finite presentation, then there exist a $t \in B-q$ and a k-morphism $f: A \to B_t$ such that $f^{-1}(q) = p$ and $f_q = r$.

F.15 Let $f: A \to B$ be a k-morphism with $f^{-1}(q) = p$. Show that if f_q is surjective then there exists $t \in B-q$ such that the map $A \to B_t$ induced by t is surjective.

F.16 Suppose A and B are of finite presentation over k and let $r: A_p \longrightarrow B_q$ be a local isomorphism. Then there exists $s \in A-p$ and $t \in B-q$ and a k-isomorphism $f: A_s \longrightarrow B_t$ such that $f^{-1}(q) = p$ and $f_q = r$ (Hint: F.14 and F.15 we can find $t' \in B-q$ and a surjective k-morphism $f: A \to B_{t'}$ inducing r. The kernel of f is finitely generated by F.4 and F.5).

II. Appendix T, tangentspaces

Let k denote a ring. For a k-module N we have the k-algebra $D_k(N)$ (see I.1.10) and the canonical k-morphism

$$\epsilon_N: D_k(N) \to k$$

If $f: N \to M$ is a k-linear map we get a k-morphism $D_k(f): D_k(N) \to D_k(M)$ by $D_k(f)(x,n) = (x, f(n))$. Note, $\epsilon_M \cdot D_k(f) = \epsilon_N$.

Let now X: {k-algebras} $\Rightarrow$ {sets} be a covariant functor, and $x \in X(k)$. For a k-module N we let $T_{X,x}(N)$ denote the fiber of

$$X(\epsilon_N): X(D_k(N)) \to X(k)$$

at x. Note that

$$N \mapsto T_{X,x}(N)$$

is a covariant functor from {k-modules} to {sets}. X is said to be infinitesimally linear at x if $T_{X,x}$ preserves finite products.

Definition T.2 Let X be infinitesimally linear at $x \in X(k)$. For a k-module N, $T_{X,x}(N)$ will be considered a k-module via the k-module structure induced from N. $T_{X,x}(k)$ will be denoted $T_x(X)$ and is called the tangent space of X at x.

Proposition T.3 Let k denote a ring, A a k-algebra. $X = \mathrm{Hom}_k(A,-)$

$$X: \{k\text{-algebras}\} \Longrightarrow \{sets\}$$

Let $x \in X(k)$. For a k-module N, let N_x denote the A-module obtained by restricting scalars along $x: A \to k$. The two functors from {k-modules} to {sets}

$$N \mapsto T_{X,x}(N)$$

and

$$N \mapsto \mathrm{Der}_k(A, N_x)$$

are isomorphic.

Proof: Straight forward.

Cor. T.4 Let $X = \mathrm{Hom}_k(A,-)$, where A is a k-algebra. X is infinitesimally linear at any point $x \in X(k)$, and

$$T_x(X) \simeq \mathrm{Der}_k(A, k_x)$$

Proof: Straight forward.

Exercise $k = \mathbb{R}$. Define X by

$$X(k') = \{(x,y) \in k'^2 \mid x^5 - y^2 - 5 = o\}$$

Find the dimension of $T_P(X)$ at all $P \in X(\mathbb{R})$.

The same for $\quad x_1^2 + x_2^2 + \ldots + x_n^2 = 1$

Exercise Let k denote a field. Define

$$S\ell_{n,k} \text{ and } O_{n,k}: \{k\text{-algebras}\} \Longrightarrow \{sets\}$$

by

$$S\ell_{n,k}(k') = \{R \in M_n(k') \mid \det R = 1\}$$

$$O_{n,k}(k') = \{R \in M_n(k') \mid {}^tR R = E\}$$

Find the tangenspace of $S\ell_{n,k}$ and $O_{n,k}$ at the origin.

Exercise Let X be a covariant functor: {k-algebras} $\Longrightarrow$ {groups}
Assume that X is infinitesimally linear at the origin, $e \in X(k)$. Show that the kernel of

$$X(D(k)) \to X(k)$$

is abelian and is identical with the underlying abelian group of $T_e(X)$.

T.5 The map induced on the tangent spaces by a morphism

Let X, Y be covariant functors

$$\{k\text{-rings}\} \Longrightarrow \{sets\}$$

and $f: X \to Y$ a natural transformation. Let $x \in X(k)$ and $y = f(k)(x)$. $D_k(k) \to k$ induces a commutative square

$$\begin{array}{ccc} X(D_k(k)) & \longrightarrow & Y(D_k(k)) \\ \downarrow & & \downarrow \\ X(k) & \longrightarrow & Y(k) \\ x & \longmapsto & y \end{array}$$

and whence a map

$$T_x(f): T_x(X) \to T_y(Y)$$

the tangent map of f at x. If X,Y are inf. linear at x,y, then $T_x(f)$ is k-linear.

<u>Exercise</u> Let k denote a ring, $n \in \mathbb{N}$. Let $G\ell_{n,k}$ denote $k' \mapsto G\ell_n(k')$. Compute the tangent map at the origin of

$$\det: G\ell_{n,k} \to G\ell_{1,k} .$$

T.6 Fiber of a morphism

<u>Remark</u> Let X: {k-algebras} $\Rightarrow$ {sets} be a covariant functor which is infinitesimally linear at $x \in X(k)$

For a k-module N, let $x(N)$ denote the image of x by $X(k) \to X(D_k(N))$. The image of $x(N)$ by $X(D_k(N)) \to X(k)$ is x, whence we can view $x(N)$ as an element of $T_{X,x}(N)$. In fact

$$x(N) = 0$$

<u>Proof</u>: Let $f: M \to N$ be a k-linear map. The commutative diagram

$$\begin{array}{ccc} & X(k) & \\ \nearrow & & \nwarrow \\ X(D_k(M)) & \xrightarrow{X(D_k(f))} & X(D_k(N)) \end{array}$$

shows that

$$T_{X,x}(f)(x(M)) = x(N)$$

Take $M = 0$ Q.E.D.

<u>Definition</u> Let Y: {k-alg.} $\Rightarrow$ {sets}, and $y \in Y(k)$. For a k-algebra k', y/k' denotes the image of y by the canonical map $Y(k) \to Y(k')$. - Let X,Y be contravariant functors

$$\{k\text{-alg.}\} \Rightarrow \{\text{sets}\}$$

and $f: X \to Y$ a natural transformation. For $y \in Y(k)$ define <u>the fiber, X_y of f at y</u> as follows: $X_y(k')$ = the fiber of $f(k'): X(k') \to Y(k')$ at y/k'.

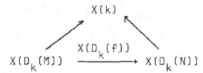

<u>Lemma</u> Let $f\colon X \to Y$ be a transformation as above. If X is infinitesimally linear at $x \in X(k)$ and Y inf. linear at $y = f(x)$, then X_y is infinitesimally linear at x and the following sequence of k-modules is exact

$$0 \to T_x(X_y) \to T_x(X) \xrightarrow{T_x(f)} T_y(Y).$$

<u>Proof</u>: Consider a k-module N. We have the following commutative diagram

$$\begin{array}{ccccc} X_y(D_k(N)) & \longrightarrow & X(D_k(N)) & \longrightarrow & Y(D_k(N)) \\ \downarrow & & \downarrow & & \downarrow \\ X_y(k) & \longrightarrow & X(k) & \longrightarrow & Y(k) \\ x & & x & & y \end{array}$$

By the previous remark $y_{/D_k(N)} = 0$, whence

$$T_{X_y,x}(N) \xrightarrow{\simeq} \mathrm{Ker}(T_{X,x}(N) \to T_{Y,y}(N))$$

The expression on the right hand side is easily seen to be product preserving Q.E.D.

<u>Proposition</u> If k is a ring, A and B are k-algebras, $X = \mathrm{Hom}_k(A,-)$, $Y = \mathrm{Hom}_k(B,-)$, $f\colon X \to Y$ a morphism, $y \in Y(k)$. Then $X_y \simeq \mathrm{Hom}_k(A \otimes_B k,-)$, where k is viewed a B-algebra via $y\colon B \to k$.

<u>Definition</u> Let X, Y, S be covariant functors from {k-algebras} to {sets} and $f\colon X \to S$, $g\colon Y \to S$ morphism.

$$X \times_S Y\colon \{k\text{-algebras}\} \Longrightarrow \{\text{sets}\}$$

denote the functor $k' \mapsto X(k') \times_{S(k')} Y(k')$.
We have natural projections

$$\begin{array}{ccccc} & & X & & \\ & \overset{p_1}{\nearrow} & & \overset{f}{\searrow} & \\ X \times_S Y & & & & S \\ & \underset{p_2}{\searrow} & & \underset{g}{\nearrow} & \\ & & Y & & \end{array}$$

<u>Proposition</u> Let $z \in X \times_S Y(k)$, $x = p_1(z)$, $y = p_2(z)$, $s = f(x) = g(y)$. If X, Y, S are infinitesimally linear at x, y, s then $X \times_S Y$ is infinitesimally linear at z and

$$\begin{array}{ccccc} & & T_x(X) & & \\ & \nearrow & & \searrow & \\ T_z(X \times_S Y) & & & & T_s(S) \\ & \searrow & & \nearrow & \\ & & T_y(Y) & & \end{array}$$

is a fibered product.

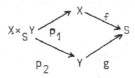

Proof: The same as the previous proof.

Exercise Let k denote a ring, V a k-module. For a k-algebra k' let $G_V(k')$ denote the set of submodules of $V\otimes_k k'$ which are direct factors in $V\otimes_k k'$. View G_V as a functor from k-algebras to sets.

1) Show that if $P \in G_V(k)$ and N is a k-module, then

$$T_{G_V,P}(N) = \mathrm{Hom}_k(P, N\otimes_k V/P)$$

and whence that G_V is infinitesimally linear at P with tangent space $\qquad T_P(G_V) = \mathrm{Hom}_k(P, V/P)$

2) Show that G_V is (formally) smooth if V is a finitely generated projective module.

3) For $m \in N$ let $G_V^m(k')$ denote the set of direct factors, U of $V\otimes_k k'$ which are of constant rank m (i.e. $\mathrm{rank}_{k'(p)}\, U\otimes_{k'} k'(p) = m$ for all prime ideals p of k'). Discuss G_V^m.

4) Compute the tangent map of

$$G_V^m \longrightarrow G^1_{\Lambda^m V}$$

$$G_V^m(k') \ni P' \longmapsto \Lambda^m P' \in G^1_{\Lambda^m V}(k')$$

Remark Let k denote a ring, A and B k-algebras. $X = \mathrm{Hom}_k(B, \;)$ $Y = \mathrm{Hom}_k(A, \;)$. Given a morphisms $f: X \longrightarrow Y$. For a geometric point $x \in X(k)$. We have

$$\underline{\mathrm{Ker}T_x(f) \simeq \mathrm{Hom}_k(\Omega_{B/A}\otimes_B k, k)}$$

(k is viewed as B-module via x)

Proof: The exact sequence

$$\Omega^1_{A/k}\otimes_A B \longrightarrow \Omega^1_{B/k} \longrightarrow \Omega^1_{B/C} \longrightarrow 0$$

give rise to the exact sequence

$$0 \longrightarrow \mathrm{Hom}_k(\Omega^1_{B/C}\otimes_B k, k) \longrightarrow \mathrm{Hom}_k(\Omega^1_{B/k}\otimes_B k, k) \longrightarrow \mathrm{Hom}_k(\Omega^1_{A/k}\otimes_A k, k)$$

But the morphism to the right is $T_x(f)$. Q.E.D.

T.7 Tangential criterion for smooth morphism

Let now A and B be finite type algebras over the algebraically closed field k. Put $Y = \mathrm{Hom}_k(A,-)$ and $X = \mathrm{Hom}_k(B,-)$ and let $f: X \longrightarrow Y$ be a morphism. Let $x \in X(k)$ and put $y = f(x) \in Y(k)$, X_y denote the fiber of f at y (X_y is equal to $\mathrm{Hom}_k(B\otimes_A k,-)$ where we have viewed k as an A-algebra via $y: A \longrightarrow k$). We have an exact sequence

$$0 \longrightarrow T_x(X_y) \longrightarrow T_x(X) \xrightarrow{T_x(f)} T_y(Y)$$

as it follows from T.3 and I.1.4.
It follows from Hilbert's Nullstellensatz that <u>f is unramified if and only if $T_x(f)$ is injective for all $x \in X(k)$</u>.
Similar if <u>X is smooth</u> over k, then it follows from the Jacobian criterion and II.3.6 that <u>f is smooth if and only if $T_x(f)$ is surjective</u> for all $x \in X(k)$.

<u>T.8 Upper semi continuity and $T_x(f)$ for a generic x</u>

Recall that an integer valued function r on a topological space Z is called upper-semi continuous if for any $x_0 \in Z$

$$r(x) \leq r(x_0)$$

in a neighbourhood of x_0.

<u>Lemma</u> Let A denote a ring, N a finitely generated A-module. For a prime ideal p of A, let k(p) denote the fraction field of A/p

$$p \mapsto rk_{k(p)} N \otimes_A k(p)$$

is upper semi-continuous.

<u>Proof</u>: Use Nakayamas Lemma. Q.E.D.

With the notation of T.7

$$X(k) \ni x \mapsto rk_k T_x(X)$$

is upper semi continuous. - Suppose X is an affine variety (i.e. the affine coordinate ring B for X is an integral domain, $X = Hom_k(B, \ )$. Then we have $rk_k(T_x(X)) = \dim X$ on a non empty open set of X(k). In general $rk_k T_x(X) \geq \dim X$. Equality if and only if X is smooth at x (compare II.6 and II.7).
Consider (still with the notation of T.7) the morphism

$$f: X \longrightarrow Y$$

It follows from the last remark in T.6, that

$$X(k) \ni x \mapsto rk_k Ker(T_x(f))$$

is upper semi-continuous.

<u>Proposition</u> Let $f: X \longrightarrow Y$ be a <u>dominant</u> morphism between affine varieties (this means that the induced map on the affine coordinate rings is injective). The fraction field of the affine coordinate ring of X is called the <u>function field</u> of X and is denoted K_X). The dimension of X is $trdg_k K_X$. We have

$$\begin{aligned} &rank_k \ Ker \ T_x(f) - rank_k \ Cok \ T_x(f) \\ &= \ \dim X - \dim Y \end{aligned}$$

for all $x \in X(k)$ in a non empty Zariski open subset of $X(k)$ ("the above formula holds for a generic $x \in X(k)$"). $T_x(f)$ is surjective for all x in a non-empty Zariski open subset of $X(k)$ if and only if $K_Y \to K_X$ is a separable field extension. $T_x(f)$ is injective (and whence bijective) for all x in a non-empty Zariski open subset of X if and only if $K_Y \to K_X$ is a finite separable extension.

Proof: Let A, resp. B denote the affine coordinate ring for X, resp. Y. Consider the A linear map

(*) $$\Omega^1_{B/k} \otimes_B A \to \Omega^1_{A/k}$$

We may and will assume that the two A-modules in (*) and the cokernel of (*) are free A-modules by the following lemma.

Lemma Let A denote an integral domain and P a finitely generated A-module. Then there exists $s \neq 0$ in A such that P_s is a free A_s-module.

Proof: Let K denote the fraction field of A. Let $p_1, \dots, p_n$ be such that $p_1 \otimes 1, \dots, p_n \otimes 1$ is a basis for $P \otimes_A K$. Let $f: A^n \to P$ be the corresponding map of A-modules. $f \otimes 1_K$ is injective, whence f is injective. Let C denote the cokernel of f. C is finitely generated and $C \otimes_A K = 0$ whence there exist $s \in A$, $s \neq 0$ such that $C_s = 0$. Q.E.D.

If we apply $\otimes_A K_X$ to (*) we get

$$\Omega^1_{K_Y/k} \otimes_{K_Y} K_X \to \Omega^1_{K_X/k}$$

by I.5.8, this complex is homotopic to $\Omega^1_{K_Y} \otimes_{K_Y} K_X \to \Omega^1_{K_X}$. Let $x \in X(k)$, and let k_x denote k viewed as an A-module via $x: A \to k$. If we apply $\mathrm{Hom}_A(\ ,k_x)$ to (*) we get

$$T_{f(x)}(Y) \leftarrow T_x(X) \leftarrow \mathrm{Ker}\, T_x(f) \leftarrow 0 .$$

Conclusion by I.4.1 , I.5.2 and I.3.3 . Q.E.D.

Gluing tangent spaces together, TX

Let X: {k-algebras} $\Rightarrow$ {sets}.
Define TX: {k-algebras} $\Rightarrow$ {sets} as follows:

For a k-algebra k',
$TX(k') = X(D_{k'}(k'))$
For a k-morphism $k' \xrightarrow{f} k''$
we have a k-morphism $D_{k'}(k') \to D_{k''}(k'')$,
this induces $TX(f): TX(k') \to TX(k'')$.

Define, the canonical projection

$$\pi: TX \longrightarrow X$$

as follows, $\pi(k'): TX(k') \longrightarrow X(k')$

is $X(\epsilon_{k'}: D_{k'}(k') \longrightarrow k')$

Proposition T.9 Let k denote a ring. $X = \mathrm{Hom}_k(A,-)$. Then

$$TX \simeq \mathrm{Hom}_k(S^{\bullet}_A \Omega^1_{A/k}, -)$$

where $S^{\bullet}_A \Omega^1_{A/k}$ is the symmetrical algebra over the A-module $\Omega^1_{A/k}$.

Proof: Straight forward.

Cor. T.10 (vector fields) Let $X = \mathrm{Hom}_k(A,-)$. The sections to the canonical projection $\pi: TX \longrightarrow X$ (i.e. natural transformations $s: X \longrightarrow TX$ such that $\pi \circ s = \mathrm{id}$) may be identified with $\mathrm{Der}_k(A,A)$.

Proposition T.11 Let k denote a field

$$G: \{k\ \text{algebras}\} \longrightarrow \{\text{groups}\}$$

a covariant functor which is representable by a k algebra A of finite type (i.e. a linear algebraic group over k). Then, we have a canonical isomorphism of A-modules

$$\Omega^1_{A/k} \simeq A \otimes_k \omega_{G/k}$$

where $\omega_{G/k}$ denotes the k-linear dual to $T_e(G)$. -

If f is a morphism of linear algebraic groups

$$f: G \longrightarrow H$$

(i.e. a natural transformation)

and $G = \mathrm{Hom}_k(B,-)$ $H = \mathrm{Hom}_k(A,-)$ $f = \mathrm{Hom}_k(t,-)$, where $t: A \longrightarrow B$ is a k-morphism, then

$$d_k(t): \Omega^1_{A/k} \otimes_A B \longrightarrow \Omega^1_{B/k}$$

is isomorphic to $B \otimes_k T_e(f)^*$ where $T_e(f)^*$ is the k-linear dual to $T_e(f): T_e(G) \longrightarrow T_e(H)$.

Proof: "G operates on G by left translation whence on TG. $G \times T_e G \longrightarrow TG$ is an isomorphism. Q.E.D."

We are going to make a series of remarks which will give sense to the above proof.

Remark T.12 Given a linear algebraic group G over k and a functor X of the form $\mathrm{Hom}_k(A,-)$. Then a (left) action of G on X is an action of $G(k')$ on $X(k')$ for all k-algebras k', subjected to the obvious functorial compatibility. It follows from the definition T.7 that if G acts on X, then G acts on TX. This action makes $TX \to X$ G-equivariant.

Remark T.13 Let k denote a field, and $X = \mathrm{Hom}_k(A,-)$ where A is a k-algebra of finite type. Let $x \in X(k)$. Define

$$T_x X: \{k\text{-algebras}\} \Rightarrow \{\text{sets}\}$$

by $T_x X(k')$ = the fiber of

$$X(D_{k'}(k')) \to X(k')$$

at x', where x' is the image of x by $X(k) \to X(k')$. Note

$$T_x X(k) = T_x(X).$$

It follows from T.3 that

$$T_x X(k') \xrightarrow{\sim} T_x(X) \otimes_k k'.$$

Note that we have a natural imbedding $T_x X \to TX$.

Remark T.14 Let $X = \mathrm{Hom}_k(A,-)$. Put G_m: {k-algebras} $\Rightarrow$ {groups} where $G_m(k')$ = the group of multiplicative units in k'. It is clear from the definition of TX that G_m acts on TX.

Note if a linear algebraic group acts on X, then the action of G on TX commutes with the action of G_m on TX.

Remark T.15 Returning to the proof of T.11. Let G act on itself by left translation; the combination of the three preceeding remark yields a transformation

$$\begin{array}{ccc} G \times T_e G & \longrightarrow & TG \\ & {\scriptstyle p_1}\searrow \quad \swarrow{\scriptstyle \pi} & \\ & G & \end{array}$$

The diagram above is commutative, and $G \times T_e G \to TG$ is an isomorphism. Let G_m act on $G \times T_e G$ via second factor.
$G \times T_e G \to TG$ is G_m-equivariant. Note, that

$G \times T_e(G)$ is represented by

$$A \otimes_k S^{\bullet}_k \omega_{G/k} \simeq S^{\bullet}_A(A \otimes_k \omega_{G/k})$$

and TG by $S^{\bullet}_A(\Omega^1_{A/k})$

Whence by Yoneda

$$S^{\bullet}_A(\Omega^1_{A/k}) \text{ and } S^{\bullet}_A(A \otimes_k \omega_{G/k})$$

are A-isomorphic. Conclusion by the next two remarks.

Remark T.16 (vector bundles) Let k denote a ring, and A a k-algebra. $X = \mathrm{Hom}_k(A,-)$. For an A-module N, put $V(N) = \mathrm{Hom}_k(S^{\bullet}_A(N),-)$. We have a canonical projection

$$\begin{array}{c} V(N) \\ \downarrow \\ X \end{array}$$

Note, that we have a natural action of G_m on $V(N)$, which makes the canonical projection equivariant, when we let G_m act trivially on X.

A triple $$(V,\ p,\ a)$$
where V is a covariant functor from {k-algebras} $\Rightarrow$ {sets}, $p: V \to X$ a natural transformation, and a an action of G_m on V making $p: V \to X$ equivariant (trivial action on X) is called a G_m-space above X. A morphism of G_m-space above X is a transformation which commutes with projections onto X and G_m-actions.

$$N \mapsto V(N)$$

is in a natural way a contravariant functor from {A-modules} to {G_m-spaces above X}

<u>Lemma T.17</u> With the notation of T.16, the functor V: {A-modules} $\Rightarrow$ {G_m-spaces above X} is fully faithful.

<u>Proof</u>: A morphism from $V(N)$ to $V(M)$ will give (by Yoneda) an A-algebra morphism $S_A^{\bullet}M \to S_A^{\bullet}N$, the grading is preserved by the following remark.

<u>Remark T.18</u> Let A denote a **Z**-graded k-algebra. $G_m(k')$ acts on $X(k') = \mathrm{Hom}_k(A,k')$ by the formula

$$ux(a) = \sum_n x(u^n a_n)$$

where $u \in G_m(k')$, $x \in X(k')$, $a \in A$ and $a = \sum_n a_n$ is the decomposition of a in homogenuous components. This is easily seen to define a G_m-action on X, and if $A \xrightarrow{f} B$ is a morphism of graded k-algebras, then the corresponding transformation

$$Y \to X \quad (Y = \mathrm{Hom}_k(B,-))$$

is G_m - equivariant.

<u>T.19</u> The above functor is an (anti) equivalence between the category of **Z**-graded algebras and the category of representable functors

$$X: \{k\text{-algebras}\} \Rightarrow \{sets\}$$

equipped with a G_m-action, the morphisms being G_m-equivariant transformations.

<u>Proof</u>: (scetch) G_m is represented by $k[T,T^{-1}]$. Let G_m act on $X = \mathrm{Hom}_k(A,-)$, and let the action map $G_m \times X \to X$ be represented by

$$c: A \longrightarrow k[T,T^{-1}]\otimes_k A = A[T,T^{-1}]$$

Define k linear maps $c_n : A \longrightarrow A$ $(n\in\mathbb{Z})$ by $c(a) = \sum_{n\in\mathbb{Z}} c_n(a)T^n$, $a\in A$

Explicit the action-axioms in terms of $(c_n)_{n\in\mathbb{Z}}$ to see that they are orthogonal projections for a structure of graded k-algebra on A.

Q.E.D.

Exercise T.20 Let k denote a field A, an associative k-algebra. $\beta: A \longrightarrow A$ a k-linear automorphism with

1) $\beta^2 = id$

2) $\beta(ab) = \beta(b)\beta(a)$, $\beta(1) = 1$

Define $G_\beta : \{k\text{-algebras}\} \Longrightarrow \{groups\}$

by $G_\beta(k') = \{x\in A\otimes_k k' \mid \beta_{k'}(x)x = 1\}$

Show that $T_e G_\beta = \{a\in A \mid \beta(a)+a = 0\}$

Show also, that G_β is smooth if

$$A \xrightarrow{\beta+id} A \xrightarrow{\beta-id} A$$

is exact.

Chapter III ETALE MORPHISMS

Definition 1.1 A morphism $A \to B$ is called etale if it is formally etale and essentially of finite presentation.

Thus etale morphisms are as well unramified as smooth. The study of etale morphisms is based on Zariski's Main Theorem and begins in III. 2. Here we shall just collect a few basic facts (compare II. F).

1.2 If S is a multiplicative subset of the ring A, then $A \to S^{-1}A$ is etale.

1.3 The composition of two etale morphisms is etale.

1.4 If $A \to B$ is etale and $A \to A'$ is any morphism then $A' \to B \otimes_A A'$ is etale.

1.5 If $A \xrightarrow{f} B \xrightarrow{g} C$ are morphisms then f unramified and gof etale implies g etale.

1.6 If $A \to B$ is essentially of finite presentation and $A \to A'$ is faithfully flat, then $A' \to B \otimes_A A'$ etale $\Rightarrow$ $A \to B$ etale.

1.7 If $f: A \to B$ is essentially of finite presentation, and $A_{f^{-1}(m)} \to B_m$ is etale for all maximal ideals m of B, then f is etale.

1.8 If $f: A \to B$ is essentially of finite presentation and q is a prime ideal of B such that $A_{f^{-1}(q)} \to B_q$ is etale, then there exist $t \in B - q$ such that $A \to B_t$ is etale.

1.9 Let k denote a field. A k-algebra A is etale if and only if it is a finite direct product of finite separable field extensions of k.

1.10 Let $f: A \to B$ be a morphism essentially of finite presentation. f is etale if and only if $\check{T}.(f)$ is an isomorphism.

1.11 Let k denote an algebraically closed field $f: A \to B$ a morphism of finite type k-algebras, $x: B \to k$ a geometric point of B and $y = f(x) (= x \circ f)$. If B is smooth over k, then $A_y \to B_x$ is etale if and only if the tangent map $T_x(f)$ is an isomorphism.

III.2 Local structure theorem

Theorem 2.1 An unramified local morphism $O \to O''$ can be factored

$$O \to O' \to O''$$

where $O' \to O''$ is a surjective local morphism and O' is of the form

$$(O[X]/(P))_q$$

where

i) $P \in O[X]$ is monic.

ii) q is a prime ideal in $O[X]$ containing P and contracting to the maximal ideal in O.

iii) $P' \notin q$.

Proof Let k denote the residue field of O. We shall make use of the fact that a finite separable field extension of k is of the form $k[X]/P$ where P is irreducible and $P' \neq 0$, and of the following version of Zariski's Main Theorem (see IV. 2.2).

Z.M.T. Let $O \to O''$ be a local morphism essentially of finite type and let k denote the residue field of O. If $O'' \otimes_O k$ is finite, then O'' is essentially finite (i.e. of the form D_p, where D is finite over O and p is a prime ideal lying over the maximal ideal of O).

Choose D and p as in ZMT. Decompose $D \otimes_O k$ in local factors $A_1 \times \ldots \times A_n$, where A_1 corresponds to p. A_1 may be identified with the residue field of O''. Choose $x \in D$ such that the projection of $x \otimes 1$ on A_1 generates that separable field extension and the projection on A_i $(i \geq 2)$ is zero. Let C denote the subalgebra of D generated by x and let r denote the restriction of p to C. Note that p is the only prime ideal of D which lies over r, whence $C_r \to D_p$ is an isomorphism ($D \otimes_C C_r$ is finite over C_r and local whence equals D_p, conclusion by Nakayama).

Whence we may assume that D is generated by a single element z. Let n denote the rank of $D \otimes_O k$. By Nakayamas lemma $1, z, \ldots, z^{n-1}$ generates D. Whence we can find $P \in O[X]$ monic of degree n and a surjection

$$O[X]/P \to D$$

which is an isomorphism after tensoring with k. Let $q \subseteq O[X]$ denote the inverse image of p.

$$(O[X]/P)_q \to D_p \quad (= O'')$$

is surjective and is an isomorphism after tensoring with k, consequently $P' \notin q$ since $O'' \otimes_O k$ is unramified over k. Q.E.D.

Cor. 2.2 Any etale local morphism $O \to O''$ is of the standard form 2.1.

Proof Let $O \to O' \to O''$ factor $O \to O''$ as in 2.1. We claim that $O' \to O''$ is an isomorphism. Let I denote the kernel of $O' \to O''$. I is finitely generated by II. F.11, and $\check{T}.(O''/O)$ may be represented by

$$I/I^2 \to o$$

according to II.3.5. Whence $I/I^2 = o$ by 1.10 and thus $I = o$ by Nakayamas Lemma. Q.E.D.

Cor. 2.3 An unramified morphism is etale if and only if it is flat.

Proof Let us first prove the corollary for an unramified local morphism $O \to O''$. Factor $O \to O''$ as in 2.1:

$$O \to O' \to O''$$

assume O'' flat over O. Then $O' \to O''$ is flat by I.2.7, $O' \to O''$ is faithfully flat by Bourbaki, Alg. Comm. I, §3, n^o5, Prop. 9, but a faithfully flat morphism is injective by loc. cit. - Conversely, if $O \to O''$ is etale then it is flat by 2.2.

The general case follows easily from 1.6 and loc. cit. II, §3, n^o4, Prop. 15. Q.E.D.

III.3 Applications to smooth morphisms

Proposition 3.1 A smooth local morphism $O \to O''$ can be factored $O \to O' \to O''$ where $O' \to O''$ is an etale local morphism and $O' = O[X_1,\ldots,X_n]_q$ where q is a prime ideal of O contracting to the maximal ideal of O.

Proof Let $x_1,\ldots,x_n$ be elements of O'' such that $d(x_1),\ldots,d(x_n)$ form a basis for $\Omega^1_{O''/O}$. Consider the O-morphism $O[X_1,\ldots,X_n] \to O''$ given by $X_i \mapsto x_i$. Let the maximal ideal contract to the prime ideal q of $O[X_i,\ldots,X_n]$ and let O' denote the local ring of $O[X_i,\ldots,X_n]$ at q.

$$\Omega^1_{O'/O} \otimes_{O'} O'' \to \Omega^1_{O''/O}$$

is an isomorphism by construction. Since O''/O is smooth, the above map represents $\check{T}.(O''/O')$ by II. 3.6, $O' \to O''$ is etale by 1.10. Q.E.D.

Cor. 3.2 A smooth morphism is flat.

Proof Let $f: A \to B$ be a smooth morphism. For a prime ideal q of B, the morphism

$$A_{f^{-1}(q)} \to B_q$$

is smooth and whence flat by 3.1 and 2.3. Flatness is local, Bourbaki, Alg. Comm. II, §3, n^o4, Prop. 15. Q.E.D.

Let $f: A \to B$ be a morphism and p a prime ideal of A, $k(p)$ denotes the fraction field of A/p. $B\otimes_A\overline{k(p)}$ is called the geometric fiber of f at p.

Proposition 3.3 Let $f: A \to B$ be a morphism essentially of finite presentation. f is smooth if and only if f is flat and all geometric fibers of f are regular (i.e. all local rings of the geometric fibers are regular local rings).

Proof Assume f is flat with regular geometric fibers. By II, 4.5 and II. 4.8 it suffices to prove that

$$\check{T}.(f)\otimes_B k(q)$$

is injective for all primes q of B. Put $p = f^{-1}(q)$. By II. 6.3 $B\otimes_A\overline{k(p)}$ is smooth over $\overline{k(p)}$ and hence by II. 4.9, $B\otimes_A k(p)$ is smooth over k(p). Whence

$$\check{T}.(B\otimes_k k(p)/k(p))\otimes_{B\otimes_k k(p)} k(q)$$

is injective by II. 3.4.

Since f is flat, we have according to II. 3.7, c.

$$\check{T}.(f)\otimes_A k(p) = \check{T}.(B\otimes_A k(p)/k(p))$$

Note finally

$\check{T}.(f)\otimes_B k(q) = (\check{T}.(f)\otimes_A k(p))\otimes_{B\otimes_A k(p)} k(q)$ Q.E.D.

Exercise 3.4 Let k denote a field and A a smooth k-algebra of finite type.

1) Show that if p is a prime ideal in A, then there exists $t\in A-p$ such that $k \to A_t$ is of the form $k \to k[X_1,\dots,X_n] \xrightarrow{f} A_t$ where f is of finite type and etale.

2) $k = k^s$, A smooth of finite type over k^s. Assume $A\neq 0$. Show that $\mathrm{Hom}_{k^s}(A,k^s)\neq\emptyset$ (Hint: a flat finite type morphism is open by IV.3.7)

3) Combine with II. 1.11 to show: let Q_1 and Q_2 be nondegenerate quadratic forms in n variables over k. Then Q_1 and Q_2 are equivalent over a finite separable extension of k.

III.4 Grothendieck-Hensel Lemma

Definition 4.1 Let O denote a local ring. An etale neighbourhood of O is an etale local morphism $O \to O'$ which induces an isomorphism on the residue fields. We shall always identify the residue field of an etale neighbourhood of O with the residue field of O.

Theorem 4.2 Let O be a local ring with residue field k and

$$X : \{O\text{-algebras}\} \Longrightarrow \{\text{sets}\}$$

a covariant functor which is formally smooth, representable by an O-algebra essentially of finite presentation. Then, for $x \in X(k)$ there exists an etale neighbourhood O' of O and $a \in X(O')$ such that $a \mapsto x$ by the canonical projection $X(O') \to X(k)$.

Proof Let the O-algebra A represent X, let $x \in X(k)$ correspond to the maximal ideal m of A and O" denote the local ring of A at m. We have more generally

Lemma 4.3 Let $O \to O''$ be a smooth local morphism which induces a finite separable extension $k \to k''$ on the residue fields. Then there exists a (finitely generated) ideal I of O" such that $O' = O''/I$ is etale over O.

Proof Put $\bar{O}'' = O'' \otimes_O k$. And let $x \mapsto \bar{x}$ denote the projection of O" onto $\bar{O}''$. Pick elements $t_1, \ldots, t_r$ of O" such that $\bar{t}_1, \ldots, \bar{t}_r$ is a (regular) system of parameters for $\bar{O}''$. Put $I = (t_1, \ldots, t_r)$. We are going to prove that $O' = O''/I$ is etale over O.

Since O"/O is smooth we may represent $\check{T}.(O'/O)$

$$I/I^2 \to \Omega^1_{O''/O} \otimes_{O''} O'.$$

By 1.10 and II.4.4 it suffices to prove that

$$I/I^2 \otimes_{O''} k'' \to \Omega^1_{O''/O} \otimes_{O''} k''$$

is an isomorphism, or that $d(t_1) \otimes 1, \ldots, d(t_r) \otimes 1$ form a basis for $\Omega^1_{O''/O} \otimes_{O''} k''$. Note that the canonical map

$$\Omega^1_{O''/O} \otimes_{O''} k'' \to \Omega^1_{\bar{O}''/k} \otimes_{\bar{O}''} k''$$

is an isomorphism. If m denotes the maximal ideal of O" we know that

$$\bar{m}/\bar{m}^2 \to \Omega^1_{\bar{O}''/k} \otimes_{\bar{O}''} k'' \text{ is an isomorphism since } k''/k$$

is separable algebraic. Q.E.D.

III.5 Henselian rings

Let O be a local ring with residue field k. If $f \in O[T]$ then $\bar{f} \in k[T]$ denotes the polynomial obtained by reducing the coefficients of f mod the maximal ideal.

Definition 5.1 A local ring O with residue field k is called henselian if whenever f is a monic polynomial with coefficients in O and $x \in k$ is such that

$$\bar{f}(x) = o, \frac{\partial \bar{f}}{\partial T}(x) \neq o$$

then there exist $a \in O$ such that $f(a) = o$ and $\bar{a} = x$.

<u>Theorem 5.2</u> Let O be a henselian local ring. Then O has no non trivial etale neighbourhoods.

<u>Proof</u> An etale neighbourhood of O has the standard form

$$O' = (O[X]/f)_q$$

by 2.2. Since the residue field extension is trivial, the prime ideal q corresponds to the kernel of $g \mapsto \bar{g}(x)$ $(O[X] \to k)$ where $x \in k$ is a <u>simple</u> root of $\bar{f}(X)$. By definition we can find $a \in O$ such that $\bar{a} = x$ and $f(a) = 0$ $X \mapsto a$ induces a local O-morphism $O' \to O$. The composit $O' \to O \to O'$ is the identity by I. 2.5. Q.E.D.

<u>Cor. 5.3</u> (Hensels Lemma) Let O be a henselian local ring and let $X : \{O\text{-algebras}\} \Longrightarrow \{sets\}$ be a covariant functor, which is formally smooth representable by an essentially finitely presented O-algebra. Then

$$X(O) \to X(k)$$

is surjective.

<u>Proof</u> Follows from 4.2 and 5.2.

<u>Cor. 5.4</u> Let O be a local ring with residue field k, $O \to O'$ an etale morphism and $O \to O''$ a local morphism. If O" is henselian, then

$$\mathrm{Hom}_O(O', O'') \to \mathrm{Hom}_k(O' \otimes_O k, O'' \otimes_O k)$$

is bijective.

<u>Proof</u> Replacing O' by $O' \otimes_O O''$ we may assume $O = O''$. The surjectivity follows from 5.3. The injectivity from I. 2.5.

<u>Cor. 5.5</u> Let O be a henselian local ring. Any finite O-algebra (i.e. finitely generated as O-module) is the direct product of finitely many local O-algebras.

<u>Proof</u> Let k denote the residue field of O and let A denote a finite O-algebra. $A \otimes_O k$ is of finite rank over k and whence decomposes in a product of local k-algebras; whence it suffices to prove that the idempotent elements in $A \otimes_O k$ can be lifted to idempotent elements in A. If A is free then this follows from Hensels Lemma 5.3 (see II. 1.8). In general we may assume A generated by one element and now we can find a

finite free O-algebra B and an O-surjection $B \to A$. An idempotent in $A \otimes_O k$ can first be lifted to $B \otimes_O k$ and then to B. Q.E.D.

Remark 5.6 5.5 is characteristic for henselian local rings. Namely, with the notation of 5.1, x defines an idempotent in $(O[X]/f) \otimes_O k$.

Cor. 5.7 Let $O \to O'$ be a local morphism which is integral. If O is henselian, then O' is henselian.

Proof It obviously suffices to prove 5.7 in case O' is finite over O. Now it follows from 5.5.

Cor. 5.8 Let O be henselian local ring, O' a finite etale O-algebra and O" an integral O-algebra. Then

$$\mathrm{Hom}_O(O',O") \to \mathrm{Hom}_k(O' \otimes_O k, O" \otimes_O k)$$

is bijective.

Proof We may assume O" finite. O" decomposes by 5.5 into finitely many local O-algebras, whence we may assume O" is local. According to 5.7 O" is henselian. Conclusion by 5.4. Q.E.D.

Proposition 5.9 Let O denote a local henselian ring. $O' \to O' \otimes_O k$ is an equivalence between finite etale O-algebras and etale k-algebras.
We need a preparation.

Lemma 5.10 Let O be a henselian local ring and let $P \in O[X]$ be a monic polynomial, $\bar{P} = rs$ a factorization of $\bar{P} \in k[X]$ in two monic polynomials r and s which has no common factors. Then there exist monic polynomials in O such that $P = RS$ and $\bar{R} = r$ $\bar{S} = s$.

Proof The decomposition

$$k[X]/P \xrightarrow{\sim} k[X]/r \times k[X]/s$$

lifts to a decomposition

$$O[X]/P \xrightarrow{\sim} C \times D$$

Let $c \in C$, resp. $d \in D$, be the projection of the residue class x of X in $O[X]/P$ onto C resp. D. According to Nakayamas lemma we can find $R, S \in O[X]$ monic of same degree as r and s such that $R(c) = 0$, $S(d) = 0$ (Let r have degree $n; 1, c, c^2, \ldots, c^{n-1}$ generates C by Nakayama). Using $x = c+d$, $R(c) = 0$, $B(d) = 0$ and $c \cdot d = 0$ one get $R(x)S(x) = 0$ and consequently

RS is divisible by P

whence equal to P since they have same degree as P and are monic.

Q.E.D.

From this Lemma and 2.2 follows

Remark 5.11 Let O be a local henselian ring with residue field k. Any finite local etale O-algebra is of the form $O[X]/(f)$ where f is monic, f is irreducible in $k[X]$ and $f' \neq o$.

Proof (of 5.9) The functor is fully faithful by 5.8. Any etale k-algebra is of the form $O' \otimes_O k$ for some finite etale O-algebra O' by 5.11 and 5.5.

Exercise 5.12 Show that a complete local ring is henselian.

Exercise 5.13 1) Let $f = X^n + a_1 X^{n-1} + \dots + a_n$ denote a polynomial with coefficients in $\mathbb{C}$, and let t_o be a simple root in f. Show by means of the implicit function theorem that there exists an analytic function r defined on an open neighbourhood U of $a. = (a_1, \dots, a_n)$ in $\mathbb{C}^n$ with values in $\mathbb{C}$ such that $r(a.) = t_o$ and for all $u. \in U$ is $r(u.)$ root in $X^n + u_1 X^{n-1} + \dots u_n$.

2) Show by means of 1) the local ring of germs of analytic functions defined in a neighbourhood of o in $\mathbb{C}^n$ with values in $\mathbb{C}$ is henselian.

3) Show that the ring of germs of r-times differentiable functions defined in a neighbourhood of o in $\mathbb{R}^n$ with values in $\mathbb{R}$ is a henselian local ring (hint: prove an analog to 1) with complex analytic replaced by real analytic).

4) Let X be a topological space, $x \in X$, show that the ring of germs of continuous functions defined in a neighbourhood of x with values in $\mathbb{R}$ is henselian (or more generally with values in a topological field $\mathbb{R}$ with an implicit function theorem, see III. R.5).

5) Let X denote a topological space, $a_1, \dots, a_n : X \to \mathbb{C}$ continuous functions. For $(t,x) \in \mathbb{C} \times X$ define

$$f(t,x) = t^n + a(x) t^{n-1} + \dots a_n(x)$$

Assume, that for all $x \in X$ $f(t,x) = 0$ has n distinct roots. Let

$V = \{(t,x)\ \mathbb{R} \times X \mid f(t,x) = 0\}$ and let $\pi : V \to X$ denote the second projection (as a cont. map). Show that $\pi : V \to X$ locally on X is of the form $U \times [1,n] \to U$ (Hint: Let F^n denote $\{(x.,t)\ \mathbb{C}^n \times \mathbb{C} \mid t^n + x_1 t^{n-1} + \dots + x_n = 0\}$ and $p: F^n \to \mathbb{C}^n$ the projection $(x.,t) \mapsto x$. Show that (π, V) is the topological pull back of (p, F^n) along a continuous map $a: X \to \mathbb{C}^n$)

III.6 Henselization Let O be a local ring. Let Etn(O) denote the category of etale neighbourhoods of O (see 4.1), the morphisms being local morphisms. Etn(O) has the following two properties

6.1 If O_1 and O_2 are etale neighbourhoods of O, then there exists at most one morphism from O_1 to O_2.

6.2 If O_1 and O_2 are etale neighbourhoods of O, then there exist an etale neighbourhood O_3 and morphisms $O_2 \to O_3$, $O_1 \to O_3$.

Proof 6.1 follows from I. 2.5. To prove 6.2, let k denote the residue field of O, which we shall identify with the residue fields of O_1 and O_2. This gives a canonical morphism

$$O_1 \otimes_O O_2 \to k$$

The localization of $O_1 \otimes_O O_2$ at the kernel of the above morphism will do the job. Q.E.D.

Definition 6.3 Let O denote a local ring

$$\lim_{\to} \mathrm{Etn}(O)$$

will be called the henselization of O and be denoted O^h.

Proposition 6.4 Let O denote a local ring. Then O^h is a henselian local ring and $O \to O^h$ is a flat local morphism with trivial residue field extension. If $O \to O'$ is a local morphism with O' henselian, then $O \to O'$ factors uniquely through $O \to O^h$.

Proof That $O \to O^h$ is flat follows from the general fact that inductive limit of flat modules is flat. That O and O^h have the same residue field k is clear from the construction. O^h is henselian: Let $f \in O^h[X]$ be monic, $x \in k$ a simple root of $\bar{f}$. The coefficients of f come from some etale neighbourhood O' of O. f has a root in $O'' = (O'[X]/f)_{X=x}$, O'' is an etale neighbourhood of O. The last assertion follows immediately from 5.4. Q.E.D.

Exercise 6.5 Let k denote an algebraically closed field, O a smooth locality over k with trivial residue extension. Show that O^h is isomorphic to the henselization of the local ring $k[X_1,\dots,X_n]_{\{X_1,\dots,X_n\}}$ where n is the dimension of O.

III.7 Newton's method

O denotes a henselian local ring with maximal ideal m and residue field k.

As a corollary to 5.2 we have: If f is a polynomium with coefficients in O and $x \in k$ is a root in $\bar{f}$ such that $\bar{f}'(x) \neq o$, then there exist $a \in O$ such that

$$f(a) = o \text{ and } \bar{a} = x .$$

<u>Proposition 7.1</u> (Newton's method) Let f be a polynomial with coefficients in O and $a \in O$ such that

$$f(a) \equiv o \bmod mf'(a)^2$$

Then there exists $b \in O$ such that

$$f(b) = o \quad \text{and} \quad a \equiv b \bmod mf'(a)$$

(If O is an integral domain and $f'(a) \neq o$, then b is unique)

<u>Proof</u> By Taylor expansion we can write

$$f(a+f'(a)T) = f(a)+f'(a)^2T+f'(a)^2T^2g(T)$$

with $g(T) \in O[T]$. Write

$$f(a) = cf'(a)^2 \quad \text{with } c \in m$$

Put $h(T) = c+T+T^2g(T)$. We have $\bar{h}(o) = o$ $\bar{h}'(o) \neq o$ whence we can find $d \in m$ such that $h(d) = o$. $b = a+f'(a)d$ solves the problem. Q.E.D.

More generally if $F = (f_1,\dots,f_r)$ is a system of polynomials in n variables with coefficients in O, and $x \in k^n$ is a solution to $F(X) = o$. Then, if the rank of $J(x)$ (J denotes the jacobian matrix of F) is r, then we can find $a \in O$ such that

$$F(a) = o \quad \text{and } \bar{a} = x .$$

We can generalize the above proposition to

<u>Proposition 7.2</u> Let $F = (f_1,\dots,f_r)$ be a system of polynomials in n variables with coefficients in O. Let J denote the jacobian matrix of F. Then if $a \in O^n$ is such that

$$F(a) \equiv o \bmod me^2$$

where e denotes one of the $r \times r$ minors of $J(a)$, then there exists

$b \in O^n$ such that

$F(b) = o$ and

$b \equiv a \bmod me$

<u>Proof</u> We will assume that e is the subdeterminant obtained from the first r variables. We can now complete $F = (f_1,\dots,f_r)$ to

$$(f_1,f_2,\dots f_r,X_{r+1},\dots,X_n) .$$

Whence we may assume $r = n$ and $e = d(a)$, $d = \det J$.

By Taylor expansion we can write (put $T = (T_1,\dots,T_n)$)

$$F(a+eT) = F(a) + eJ(a)T + e^2G(T)$$

where $G(T)$ is a vector of polynomials each beginning with terms of degree at least 2. Let J' denote the adjoint matrix to J, such that $JJ' = JJ' = dI$. We can write $F(a) = e^2c$, where $c = (c_1,\dots,c_n)$, $c_i \in m$. Substitute $e^2I = eJ(o)J'(o)$ to get

$$F(a+eT) = eJ(o)H(T)$$

where $H(T) = J'(o)c+T+J'(o)G(T)$ whence we can find $t = (t_1,\dots,t_n)$ with $t_i \in m$ such that $H(t) = o$. $b = a+et$ will do the job. Q.E.D.

Exercise 7.3 Show that 17 is a square in $\mathbb{Q}_2$ (two-adic numbers).

Lemma 7.4 Let A denote a discrete henselian valuation ring. $G(T,X.) = G(T,X_1,\dots,X_n)$ a polynomial with coefficients in A. Suppose

$$G(o,o.) = o \quad G_T'(o,o.) \neq o$$

Then there exists a nbhd U of o in A and a nbhd V of o in A^n (adic topology) such that for all $x. \in V$ there exists a unique $t \in U$ such that $G(t,x.) = o$. Moreover the hereby defined function $V \to U$ is continuous in the adic topology.

Proof Let us first remark that it suffices to prove continuity in o. Put $e = G_T'(o,o)$. We have by Taylor expansion

$$G(eT,e^2X_1,\dots,e^2X_n) =$$

$$G(o,o) + e^2T - e^2H(T,X.)$$

where H is a polynomial with coefficients in A, with $H(o,o) = o$ and $H_T'(o,o) \neq o$. Whence we may assume

$$G(T,X.) = T-H(T,X.) \qquad H(o,o) = o$$
$$H_T'(o,o) \neq o$$

1) Existence If $x. = (x_1,\dots,x_n)$ is such that $x_i \in m$, where m denotes the maximal ideal of A, then there exists precisely one $t \in m$ such that $t = H(t,x.)$.

2) Uniqueness For later use we shall give a second proof of uniqueness: Suppose $t_1,t_2 \in m$ and $t_1 \neq t_2$. Then for $x. \in m^n$

$$\frac{G(t_1,x.)-G(t_2,x.)}{t_1-t_2} = D(t_1,t_2,x.)+1$$

where D is a polynomial with coefficients in A and $D(o,o,o.) = o$. Whence $G(t_1,x.) \neq G(t_2,x.)$ $(\neq o)$.

3) Continuity Write

$$H(T,X.) = K(T,X.) - T^2P(T)$$

where $K(T,o) = o$

The equation becomes

$$(1+TP(T))T = K(T,X)$$

From which the continuity in the point o. follows. Q.E.D.

III.8 Etale algebras over a normal domain

By a normal ring is understood an integral domain which is integrally closed in its field of fractions.

Lemma 8.1 Let O be a normal local ring with fraction field K and residue field k. Any etale, local O-algebra is of the form

$$(O[X]/(P))_q$$

where

i) $P \in O[X]$ is monic.

ii) q is a prime ideal in $O[X]$ containing P and contracting to the maximal ideal in O.

iii) $P' \notin q$

iv) P is irreducible in $K[X]$.

Proof By 2.2 we know that a local, etale O-algebra is of the form

$$(O[X]/(P))_q$$

with condition i), ii) and iii) satisfied. Factor $P = Q_1, \ldots, Q_n$ in $K[X]$ with Q_i monic and irreducible. The Q_i's has coefficients in O by Bourbaki: Alg. Comm. V, §1, n°3, Prop. 11. Whence we can write

$$P = QR$$

where R and Q are monic polynomials with coefficients in O, Q is irreducible in $K[X]$ and $Q \in q$. We have

$$P' = Q'R+QR'$$

whence $Q' \notin q$. Consider the surjective, local morphism

$$(O[X]/(P))_q \to (O[X]/(Q))_q$$

the target is flat over O, the source is unramified over O, whence the morphism is flat by I.2.7, but also faithfully flat since it is local by Bourbaki, Alg. Comm. I, §3, n°4, Prop. 15, whence it is also injective (loc. cit). Q.E.D.

Proposition 8.2 Let O be a local normal domain. f: O → O' a local, unramified morphism. f is etale if and only if f is injective.

Proof Factor $f: O \longrightarrow O'' \longrightarrow O'$ where O'' is a standard etale of the form 8.1 and $O'' \longrightarrow O'$ is surjective. Suppose f is injective. Then the composit $K \longrightarrow O''\otimes_O K \longrightarrow O'\otimes_O K$ is not zero, i.e. $O'\otimes_O K \neq 0$. $O''\otimes_O K$ is the fraction field of O'', consequently $O''\otimes_O K \longrightarrow O'\otimes_O K$ is an isomorphism and whence $O'' \longrightarrow O'$ is an isomorphism. Q.E.D.

Proposition 8.3 Let O be a normal local domain and $O \longrightarrow O'$ a local etale morphism. Then O' is a normal (local) domain.

Proof The proposition follows from 8.1 and the Corollary of the following Lemma.

Lemma 8.4 (Tate) Let A be a ring, f a monic polynomial of degree n with coefficients in A. $B = A[X]/(f)$ and x the residue class of X in B. Then, there exist $b_1,\ldots,b_n \in B$ such that for all $b \in B$

$$f'(x)b = \sum_{i=1}^{n} Tr_{B/A}(b_i b)x^i$$

Proof Over B f(X) factors

(1) $$f(X) = (X-x)\sum b_i X^i, \qquad b_i \in B$$

We are going to prove that $b_0,\ldots,b_{n-1}$ satisfy the condition of the Lemma. - Differentiate (1) to get

$$f'(X) = \sum b_i X^i + (X-x)\frac{d}{dX}\sum b_i X^{-i}$$

substitute $X = x$ to get

(2) $$f'(x) = \sum b_i x^i$$

$1,x,\ldots,x^{n-1}$ form a basis for the A-module B. Let $p_0,\ldots,p_{n-1}$ be the projections after this basis. By definition

(3) $$b = \sum p_i(b)x^i \qquad \text{all } b \in B$$

and by definition of Tr

(4) $$Tr(b) = \sum p_i(bx^i) \qquad \text{all } b \in B$$

The key relation is

(5) $p_i(b) = p_{n-1}(b_i b)$ all $b \in B$, $i=0,\ldots,n-1$.

Let us grant (5) for a moment. By (4) and (5) we have

$$Tr(b) = \sum p_{n-1}(b_i bx^i) = p_{n-1}(b\sum b_i x^i)$$

and whence by (2)

(6) $$Tr(b) = p_{n-1}(f'(x)b), \qquad \text{all } b \in B.$$

By (3) and (5)

$$f'(x)b = \sum p_i(f'(x)b)x^i = \sum p_{n-1}(b_i f'(x)b)x^i$$

and by (6): $$f'(x)b = \sum \mathrm{Tr}(b_i b)x^i .$$

Proof of (5). It suffices to prove

(7) $$p_{n-1}(b_i x^j) = \delta_{ij} \quad , \quad 0 \leq i,j \leq n-1$$

Write $f(X) = \sum a_i X^i$ and expand (1) to get

(8) $$b_i = a_{i+1} + b_{i+1}x \quad , \quad 0 \leq i \leq n-2$$

(9) $$0 = a_o + b_o x \quad , \quad b_{n-1} = 1 .$$

It is clear from (9), that (7) is true for $(0,j)$, $j \geq 1$ and $(n-1,j), j \geq 0$. Multiply (8) by x^j to get

$$p_{n-1}(b_i x^j) = p_{n-1}(b_{i+1}x^{j+1}) \quad , \quad 0 \leq i,j \leq n-2 .$$

Conclusion by a simple induction. Q.E.D.

Cor. 8.5 Let A be a normal integral domain with fraction field K, f a monic polynomial with coefficients in A and irreducible in $K[X]$. If $f'(X) \neq 0$, then

$$(A[X]/(f))_{f'}$$

is a normal domain.

Proof With the notation of 8.4 the formula holds by linearity for all $b \in K[X]/(f)$. If b is integral over A, then $\mathrm{Tr}_{B/A}(b_i b) \in A$ by Bourbaki: Alg. Comm. V, §1, Cor.2 de la Prop. 17, and whence $b \in (A[X]/(f))_{f'}$. Q.E.D.

Cor. 8.6 The henselization of a local, normal domain is a normal domain.

Cor. 8.7 Let A denote a reduced ring and f a monic polynomial with coefficients in A. Then $(A[X]/f)_{f'}$ is reduced.

Proof With the notation of 8.4 it suffices to show that if $b \in B$ is nilpotent then $\mathrm{Tr}(b)$ is nilpotent in A, i.e. contained in all minimal prime ideals q. Replace A by the fraction field of A/q and use that a nilpotent endomorphism of a vector space has trace o. Q.E.D.

Let A be a local normal domain with residue field k and fraction K. Recall (Cohen-Seidenberg) that if C is a ring containing A and integral over A, then the maximal ideals of C are precisely the prime ideals in C lying over the maximal ideal of A.

Proposition 8.8 With the notation above, let B denote the normalization of A in K. For a maximal n of B, $A^h \to B_n^h$ is integral and the kernel is a minimal prime ideal of A^h. In this way is established a one to one correspondance between the maximal ideals of B (the branches of A) and the minimal prime ideals of A^h.

Proof Put $B' = A^h \otimes_A B$

1) The fiber of B above the maximal ideal of A (i.e. $B \otimes_A k$) may be identified with the fiber of B' above the maximal ideal of A^h. - Clear.
2) If n' is a maximal ideal of B', then $B'_{n'}$ is integral over A^h and isomorphic to B_n^h

Proof: Consider B as the limit of its finite sub-A-algebras. The first part then follows from 5.5. By 5.7 $B'_{n'}$ is henselian so it obviously has the universal property of B_n^h.

3) $n' \supseteq q'$ is a one to one correspondance between the maximal ideals n' in B' and the minimal primes q' in B'.

Proof: Each maximal ideal contains a minimal and conversely. By 2) and 8.6 a maximal ideal contains precisely one minimal prime. Let q' be a minimal prime ideal then B'/q' is local since all its finite sub-A^h-algebras are local by 5.5. - Note finally the following triviality.

4) The canonical map $A^h \otimes_A K \to B' \otimes_B K$ is an isomorphism.

We can now conclude the proof by applying the following Lemma twice:

Lemma 8.9 Let A denote a ring, S a multiplicative subset of A consisting of non zero divisors in A. Then the minimal ideals in $S^{-1}A$ and A corresponds 1-1.

Proof The retraction to A of a minimal prime in $S^{-1}A$ is obviously a minimal prime in A. Suppose q is minimal in A and suppose $s \in q$ is a non zero divisor. Then s/1 is a non zero divisor in A_q on the one hand and nilpotent on the other hand. Whence exist $t \in A-q$ and integer $n \geq 1$ such that $ts^n = 0$, contradiction. Q.E.D.

We can draw the following corollary to the proof of 8.8:

Cor. 8.10 With the previous notation, let L denote a finite field extension of K. Assume A is normal and let B denote the integral closure of A in L. Then the number of maximal ideals in B is finite.

Proof $B' = B\otimes_A A^h$; according to the proof of 8.8, the maximal ideals of B corresponds 1-1 to the minimal prime ideals of $B'\otimes_B L$. $B'\otimes_B L \simeq A^h\otimes_A L$ is finite over $A^h\otimes_A K$. $A^h\otimes_A K$ is the fraction field of A^h: Write $A^h\otimes_A K = \varinjlim A'\otimes_A K$ where the limit is over all etale neighbourhoods A' of A. This shows that $A^h\otimes_A K$ is integral over K. Q.E.D.

Cor. 8.11 Let A denote a local integral domain. A^h is a local domain if and only if the normalization of A in its fraction field is a local domain.

Proof By 8.7 A^h is reduced. Conclusion by 8.8. Q.E.D.

A is called unibranched if A^h is a domain.

III.9 Local ramification theory

Let R denote a local, normal domain which is henselian. K denotes the fraction field of R and k_K denotes the residue field of R.
Let L denote a finite field extension of K; the integral closure R_L of R in L is local by 5.5 and henselian by 5.7. k_L denotes the residue field of R_L. We say that L is an unramified extension of K if R_L is an unramified R-algebra. In that case R_L is etale over R by 8.2 and moreover R_L is finitely generated and free as R-module in case L is unramified by 5.4, 5.7, 5.9 and 5.11.

Proposition With the notation above

9.1 $[k_L: k_K]_s < +\infty$ and $[k_L: k_K]_s$ divides $[L: K]$

9.2 $[k_L: k_K]_s = [L: K]$ if and only if L is an unramified extension of K

9.3 If L is unramified over K, then any subextension of L is unramified over K.

9.4 L contains a largest unramified subextension L^{nr} and $[L^{nr}: K] = [k_L: k_K]_s$

Proof Let k' denote a finite separable subextension of k_L. By 5.9 we can find a finite, etale R-algebra R' such that $R' \otimes_R k_K \cong k'$. R' is local by 5.5. R' is a normal domain by 8.3 and $R \to R'$ is injective. Let K' denote the fraction field of R'. According to 5.4 we can find an R-morphism $R' \to R_L$ which on the residue fields induces the apriori embedding of k' in k_L. $R' \to R_L$ is injective (the kernel of $R' \to R_L$ and the zero ideal i R' are prime ideals in R' which both contract to the zero ideal in R).
8.2 and 5.4 permet us now to conclude 9.1, 9.2 and 9.4. 9.3 will follow from 9.6 below.

Definition 9.5 For a finite extension L of K we set

$$e(L: K) = [L: K] / [k_L: k_K]_s$$

$$\text{i.e.} \quad [L: K] = [L^{nr}: K]e(L: K)$$

Proposition 9.6 Let N be a finite extension of L. Then

$$e(N: K) = e(N: L)e(L: K)$$

Proof Follows from 9.4 and the fact that (with a slight abuse of notation

$$[k_N: k_L]_s [k_L: k_K]_s = [k_N: k]_s .$$

Remark 9.7 $L \mapsto k_L$ is an equivalence between the category of finite unramified field extensions of K and the category of finite separable extensions of k_K. In particular an unramified extension L is a Galois Extension of K if and only if k_L is a Galois extension of k_K.

Proposition 9.8 Let L be a finite Galois extension of K. For $\sigma \in \mathrm{Gal}(L/K)$, σ denotes the k_K automorphism of k_L^s induced by $\sigma(\sigma(R_L)=R_L)$. k_L^s is a Galois extension of k_K and

$$\mathrm{Gal}(L/K) \twoheadrightarrow \mathrm{Gal}(k_L^s/k_K) ,$$

is surjective, and its kernel (the inertia group) has L^{nr} as the fixed field.

Proof L^{nr} is obviously a Galois extension of K, and whence by 9.7 $k_L nr = k_L^s$ is a Galois extension. The map above factors

$$\mathrm{Gal}(L/K) \to \mathrm{Gal}(L^{nr}/K) \to \mathrm{Gal}(k_L^s : k_{K'})$$

the map to the right is an isomorphism by 9.7. The map to the left is surjective by Galois theory and its kernel has L^{nr} as fixed field by Galois theory. Q.E.D.

Proposition 9.9 (Translation) Let L_1 and L_2 be subextensions of L and assume that $L = K[L_1,L_2]$, the smallest extension containing L_1 and L_2. If L_1 is unramified over K, then L is unramified over L_2.

Proof Put $R_i = R_{L_i}$. $R_1 \otimes_R R_2$ is an unramified R_2-algebra, whence the image R_3 of $R_1 \otimes_R R_2$ in L is an unramified R_2 algebra. R_3 has L as fraction field, is unramified over R_2 and whence normal, i.e.

$$R_3 = R_L$$ Q.E.D.

Exercise 9.10 Discuss algebraic field extensions L of K for which any finite subextension is unramified, generalize the notion L^{nr} and generalize 9.7, 9.8, 9.9.

III.10 General ramification theory

Let A denote a local, normal domain with residue field k and fraction field K. A^h denotes the henselization of A and K^h the fraction field of A^h. - Let L denote a finite extension of K, let B denote the integral closure of A in L and let $q_1,\dots,q_r$ denote the maximal ideals of B (there are only finitely many by 8.10). Put $B_i = B_{q_i}$ and let L_i^h denote the fraction field of B_i^h. - We say that <u>L is unramified</u> at q_i if $A \to B_i$ is unramified.

<u>Proposition</u>

10.1 $A^h \to B_i^h$ is integral and injective. The induced map $K^h \to L_i^h$ is a finite extension.

10.2 The canonical map $L \otimes_K K^h \to \prod_{i=1}^{r} L_i^h$ is an isomorphism.

10.3 L is unramified at q_i if and only if L_i^h is an unramified extension of K^h.

<u>Proof</u> The two A-algebras B and $B \otimes_A A^h$ have the same fiber over the maximal ideal of A. Let $p_1, \ldots, p_r$ denote the primes in $B \otimes_A A^h$ labelled such that p_i contracts to q_i via the canonical map $B \to B \otimes_A A^h = B'$. $B \otimes_A A^h$ decomposes in $B_{p_i} \times \ldots \times B_{p_r}$

Consider the diagram

$$\begin{array}{ccccccc} A^h & \longrightarrow & B \otimes_A A^h & \longrightarrow & \prod B_i^h & \longrightarrow & \prod L_i^h \\ \downarrow & & \downarrow & & & \nearrow & \\ K^h & \longrightarrow & L \otimes_K K^h & & \cdots & & \end{array}$$

We may consider $\prod L_i^h$ the total ring of fractions of $B \otimes_A A^h$ whence the existence of the dotted arrow making the diagram commutative. This shows $A^h \to B_i^h$ is injective. - $L \otimes_K K^h$ is reduced since $K \to K^h$ is separable, $L \otimes_K K^h$ is of finite rank over K^h. In conclusion, $L \otimes_K K^h$ may be considered the total ring of fractions of $B \otimes_A A^h$. This proves 10.1 and 10.2.

10.3 If $A \to B_i$ is unramified then $A^h \to B_i^h$ is unramified since B_i^h isobtained from $B_i \otimes_A A^h$ by localization. Pick $b \in B$ such that the residue class of b generates the residue field of B_i and such that $b \in q_j$ for $j \neq i$. Put $C = A[b]$ and let C_i denote the localization of C at the restriction to C of q_i. $B \otimes_C C_i$ is local, whence $B \otimes_C C_i = B_i$. This implies that $C_i \to B_i$ is integral. This implies that $B_i \otimes_{C_i} C_i^h$ has the universal property of B_i^h.

We want to show that $C_i = B$. Since $B^h \simeq B \otimes_{C_i} C_i^h$ it suffices to prove that (C_i^h is faithfully flat over C_i)

$$C_i^h \to B_i^h \quad \text{is an isomorphism.}$$

We know it is injective from the description $B^h \simeq C^h \otimes_C B$, that is has trivial residue extension by choice of b, that it is finite and unramified (since B^h is finite and unramified). Conclusion by Nakayamas Lemma. Q.E.D.

Remark It follows from the proof above that $B \otimes_A A^h \simeq \prod B_i^h$. Whence, if L is unramified at all q_i's, then B is a finitely generated free A-module.

Remark 10.4 (1st principle of algebraic number theory) The fiber of B above the maximal ideal of A can be recovered from $L \otimes_K K^h$, Namely, decompose $L \otimes_K K^h \simeq \prod L_i'$ and take the integral closure B_i' of A^h in L_i'. The fiber of $\prod B_i'$ above the maximal ideal of A^h may be identified with the fiber of B above the maximal ideal of A.

Example 10.5 $K = \mathbb{Q}$ $L = \mathbb{Q}(\sqrt{17})$

The fiber above

2	: $\mathbb{F}_2 \times \mathbb{F}_2$
17	: $D_{\mathbb{F}_{17}}$ (dual numbers)
$p \neq 2, 17$, and 17 a square mod p	: $\mathbb{F}_p \times \mathbb{F}_p$
"-" 17 not "-"	: $\mathbb{F}_{p^2}$

17 is square in $\mathbb{Q}_2^h$ by Newtons method (see 7.3).

17 is not a square in $\mathbb{Q}_{17}^h$ since $\mathbb{Z}_{17}^h$ is a discrete valuation ring with 17 as a parameter, whence the second result.

If 17 is a square mod p $(p \neq 2, 17)$ then 17 is a square in $\mathbb{Q}_p^h$ by Hensel's Lemma.

If 17 is not a square mod p then $\mathbb{Z}_p^h[X]/X^2-17$ is etale and whence normal. Q.E.D.

Cor. 10.6 Let N be a subextension of L. If L is unramified at q_i then N is unramified at the restriction of q_i to N.

Proof Follows from 10.3 and 9.3.

Remark 10.7 Let L_1 and L_2 be subextensions of L such that $L = K[L_1, L_2]$. If L_1 is unramified at q_i, then L is unramified over L_2 at q_i. - The proof is the same as that of 9.9.

Definition 10.8 Let k_i denote the residue field of B_i and k the residue field of A. Put $f_i(L: K) = [k_i : k]_s$ which is finite by the proof of 10.1 and 9.1. Put $e_i(L: K) = e(L_i^h : K^h)$.

Cor.

10.9 If N is a subextension of L, then

$$e_i(L: K) = e_i(L: N)e_i(N: K)$$
$$f\ (L: K) = f\ (L: N)f\ (N: K)$$

10.10 $e_i(L: K) = 1$ if and only if L is unramified over K at q_i

10.11 $\sum_{i=1}^{r} e_i(L: K)f_i(L: K) = [L: K]$

Proposition 10.12 Suppose L is a (finite) Galois extension of K with Galois group G. Put $G_i^D = \{\sigma\in G \mid \sigma(q_i) = q_i\}$ (the decomposition group for q_i) and $G_i^I = \{\sigma\in G^D \mid \sigma$ operates trivially on $k_i^s\}$ (the inertia group for q_i). Then, L_i^h is a Galois-extension with Galois group G_i^D. k_i^s is a Galois extension with Galois group G_i^D/G_i^I. The fixed field for G_i^D is the largest sub-extension of K, unramified at q_i and having trivial residue extension at q_i. The fixed field for G_i^D is the largest sub-extension of L at which q_i is unramified. The order of $G_i^I = e_i(L: K)$.

Proof G operates transitively on $q_1,\dots,q_r$, by Bourbaki: Alg. Comm. §2, n°3, Prop. 6, whence $r\cdot[G:G_i^D] = [L:K]$ and the L_i^h's are isomorphic, 10.2 implies that $[G:G_i^D] = [L_i^h:K^h]$. On the other hand G_i^D operates faithfully on L_i^h. This proves that L_i^h/K^h is Galois with Galois group G_i^D. Let L_i^D denote the fixed field for G_i^D. By the same argument as above $L_i^h/L_i^{D,h}$ is a Galois-extension with Galois group G_i^D. Whence $L_i^{D,h} = K^h$ and whence L_i^D is unramified over K at q_i by 10.3 and k_i^D / k is trivial. By Galois theory the sub-extension of L/L_i^D and $L_i^h/L_i^{D,h}$ are in one to one correspondance and are simultaneous unramified. We can now conclude the proof by quoting 9.8. Q.E.D.

III.R Topology on rational points

This appendix treats the local topological structure of unramified, smooth and etale morphisms defined over a topological field R with an implicit function theorem ($\mathbb{C}$, $\mathbb{R}$, $\mathbb{Q}_p$,...). For a precise definition see R.7).

Let us first make a geometrical setup.

Definition R.1 An affine R-scheme is a covariant functor

$$X: \{R\text{-algebras}\} \Longrightarrow \{\text{sets}\}$$

which is isomorphic to a functor of type $Hom_R(A,-)$, where A is an R-algebra of finite type. We call A the affine coordinate ring for X.
A morphism f: $X \to Y$ of affine R-schemes is a natural transformation of functors. By the Yoneda principle the category of affine R-schemes is dual to the category of finite type R-algebras.

A morphism $f: X \to Y$ is called a closed immersion if the induced map on the affine coordinate rings is surjective. f is called an open immersion if (let $A \xleftarrow{v} B$ denote the induced map on the affine coordinate rings) there exists $a_1,\dots,a_n \in B$ such that

i) $B_{a_i} \to A_{v(a_i)}$ is an isomorphism, all i.

ii) $v(a_1),\dots,v(a_n)$ generates the ideal A.

Let the affine R-scheme X have the affine coordinate ring $R[X_1,\dots,X_n]/(f_1,\dots,f_r)$. Identify X(R) with

$$V(f_\bullet) = \{x_\bullet \in R^n \mid f_i(x_\bullet) = 0 \text{ all } i\}$$

Give R^n the product topology and $V(f_\bullet)$ the subspace topology.

Lemma R.2 The topology on X(R) is independant of the presentation of the affine coordinate ring for X.

Proof Easy. Left to the reader.

Definition R.3 The above topology on X(R) will be called the analytical topology (to distinguish it from the Zariski topology)

Proposition R.4 The functor

$$X \longmapsto X(R)$$

from affine R-schemes to topological spaces preserves open immersions, closed immersions and fibered products.

Proof (Hint) The preservation of immersions and products is clear. For fibered products use that if $X \to Z$ and $Y \to Z$ are morphisms of affine R-schemes, then

$$X \times_Z Y \longrightarrow X \times Y$$

is a closed immersion. Q.E.D.

Definition R.5 Let $f: X \to Y$ be a morphism of affine R-schemes and $x \in X(R)$, $y = f(R)(x) \in Y(R)$. Let A and B be affine coordinate rings for X and Y respectively and let f induce the R-morphism v: B ⟶ A. View x and y as R-morphisms A ⟶ R and B ⟶ R. Let $O_{X,y}$ and $O_{Y,y}$ denote the local ring of A at (the kernel of) x, resp. the local ring at (the kernel of) y. The induced local morphism

$$O_{Y,f(x)} \longrightarrow O_{X,x}$$

we denote f_x. We shall say that <u>f is unramified, smooth, etale at x∈X(R)</u> if f_x is unramified, smooth, etale.

<u>Example R.6</u> Define affine R-schemes $\mathbb{A}^n$ and $\mathbb{E}^n$ by

$$\mathbb{A}^n(k) = k^n \quad (k \text{ an R-algebra})$$

$$\mathbb{E}^n(k) = \{(x_1,\dots,x_n,t)\in k^{n+1} \mid$$

$$P(x.,t) = t^n + \sum_{i=1}^{n} x_i t^{n-i} = 0 \text{ and}$$

$$\frac{\partial P}{\partial t}(x.,t) \text{ is a unit in } k\}$$

Note that the canonical projection π: $\mathbb{E}^n \longrightarrow \mathbb{A}^n$ $((x.,t) \longrightarrow x.)$ is etale.

<u>Definition R.7</u> The topological field R is said to be a <u>topological field with an implicit function theorem</u> if for all $n\geq 1$

$$\begin{array}{l}\mathbb{E}^n(R)\\ \downarrow\\ \mathbb{A}^n(R)\end{array}$$ is a local homeomorphism at all points of $\mathbb{E}^n(R)$ in the <u>analytical topology</u>.

(A continuous map f: S ⟶ T is called a local homeomorphism at s∈S if there exists an open nbhd U of s which is mapped homeomorphically onto an open nbhd of f(s))

From now on we shall assume that R is a topological field with an implicit function theorem.

<u>Theorem R.8</u> Let f: X ⟶ Y be a morphism of (affine) R-schemes. If f is etale at the point x∈X(R) then f: X(R) ⟶ Y(R) is a local homeomorphism at x in the analytical topology.

<u>Proof</u> According to III.2.2 (and II.F.16) we can find a morphism a: Y ⟶ $\mathbb{A}^n$ and a commutative diagram

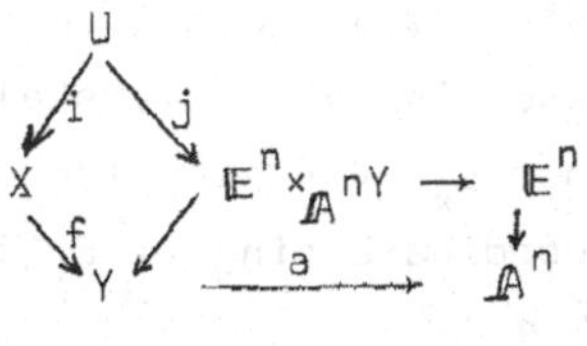

where i and j are open immersions such that x belongs to the image of U(R). Whence it suffices to prove the theorem for the case f equals the canonical projection of $\mathbb{E}^n \times_{\mathbb{A}^n} Y$ onto Y. But the diagram

$$\begin{array}{ccc} \mathbb{E}^n \times_{\mathbb{A}^n} Y(R) & \longrightarrow & \mathbb{E}^n(R) \\ \downarrow & & \downarrow \\ Y(R) & \longrightarrow & \mathbb{A}^n(R) \end{array}$$

is cartesian in the analytic topology, and the pull back of a local homeomorphism is a local homeomorphism. Q.E.D.

<u>Cor.R.9</u> Let f: X ⟶ Y be a morphism of (affine) R-schemes which is unramified in x∈X(R). Let s and t be continuous section to f_R: X(R) → Y(R) in the analytic topology. Let y = f(x). If s(y) = t(y) = x, then t = s is an analytic nbhd of y.

<u>Proof</u> By III.2.1 (and II.F.14, 15) we can find a commutative diagram

$$\begin{array}{ccccc} & & V & & \\ & {}^{j}\swarrow & & \searrow^{i} & \\ E & & & & X \\ & {}_{g}\searrow & & \swarrow_{f} & \\ & & Y & & \end{array}$$

where i is an open immersion such that x belongs to the image of V(R), j is a closed immersion and g is etale.

<u>Cor. R.10</u> Let f: X ⟶ Y be a morphism of (affine) R-schemes, and x∈X(R) a point where f is smooth. Then, f_R: X(R) ⟶ Y(R) is open at x in the analytic topology, and f_R has a continuous section s in a nbhd of y = f_R(x), with s(y) = x.

<u>Proof</u> According to III.3.1 (and II.F.16) we can find a commutative diagram

$$\begin{array}{ccccc} & & V & & \\ & {}^{g}\swarrow & & \searrow^{i} & \\ Y\times\mathbb{A}^n & & & & X \\ & {}_{p_1}\searrow & & \swarrow_{f} & \\ & & Y & & \end{array}$$

where i is an open immersion such that x belongs to the image of V(R) and g is etale. Q.E.D.

A morphism f: E ⟶ X of affine R-schemes is called <u>a flat covering of X</u> if the affine coordinate ring of E is a finitely generated and projective module over the affine coordinate ring of X. We shall say that E is unramified over x∈X(R) if the fiber E_x of f at x (see II.T.6) is unramified over R, i.e. the affine coordinate ring of E_x is an unramified

R-algebra.

Theorem R.11 (Covering space theorem) Let $f: E \to X$ be a flat covering of X and x a point of $X(R)$ where the fiber E_x is unramified. Then

1) in the analytic topology $E(R)$ is a covering space of $X(R)$ in a nbhd of x.

2) E_z is isomorphic to E_x for all $z \in X(R)$ in a nbhd of x.

Let us first give a geometric formulation to some of the result of III.

Lemma R.12 Let $E \to X$ and $F \to X$ be flat coverings of X and $x \in X(R)$ such that E and F have isomorphic and unramified fibers at x. Then there exists an affine R-scheme X', an etale morphism $\pi: X' \to X$ and $x' \in X'(R)$ with $\pi(x') = x$ such that the pull backs of E and F along π are isomorphic over X'.

Proof Let X,F and F have affine coordinate rings O,A and B. Put $O_x = O_{X,x}$, $A_x = A \otimes_O O_x$, $B_x = B \otimes_O O_x$. A_x and B_x are finite and etale over O_x: By III.2.3 it suffices to prove that $\Omega^1_{A_x/O_x} = 0$. Let R_x denote the residue field of O_x. $\Omega^1_{A_x/O_x} \otimes_{O_x} R_x$ is zero by assumption, $\Omega^1_{A_x/O_x}$ is a finitely generated O_x-module, since A_x is a finite O_x-module; conclusion by Nakayamas Lemma. - By III.5.9 A_x and B_x become isomorphic after the base extension $O_x \to O_x^h$, whence we can find an etale neighbourhood O'_x of O_x such that A_x and B_x become isomorphic after the base extension $O_x \to O'_x$. Let $(X';\pi,x')$ be a triple, where X' is an affine R-scheme $\pi: X' \to X$ an etale morphism $x' \in X'(R)$ with $\pi(x') = x$ and such that $O_{X,x} \to O_{X',x'}$ equals $O_x \to O'_x$. Modifying X' slightly (as in II.F.14-16) we may assume that the pull back of E and F along π are isomorphic over X'. Q.E.D.

Proof of R.11 Apply R.12 to the two flat coverings $f: E \to X$ and $p_1: X \times E_x \to X$. The first part now results immediately from Theorem R.8. Note, with the notation of R.12, that for $y' \in X'(R)$, the fiber of the pull back of E along π is R-isomorphic to the fiber of E at $y = \pi(y')$. Q.E.D.

Cor. R.13 Let $f = T^n + a_1T^{n-1}+\ldots+a_n$ be a polynomial with coefficients in R, having n-distinct roots in $\bar{R}$. Then for a polynomial $g = T^n + b_1T^{n-1}+\ldots+b_n$ the R-algebras $R[T]/(f)$ and $R[T]/(g)$ are isomorphic when the coefficients of g are sufficiently close to the coefficients of f.

Proof Define the affine R-scheme $\mathbb{F}^n$ as follows. For an R-algebra k put

$$\mathbb{F}^n(k) = \{(x.,t)\in k^{n+1} \mid t^n+x_1t^{n-1}+\ldots+x_n=0\}$$

$(x.,t) \to x.$ defines a flat covering $\pi: \mathbb{F}^n \to \mathbb{A}^n$. Apply R.11 to the point $(a_1,\ldots,a_n)\in \mathbb{A}^n(R)$. Q.E.D.

Example Let (E,π) be a flat covering of $\mathbb{A}^1_{\mathbb{R}} = \mathbb{R}$. If there are 5 real points in E_x for all $-1<x<0$ and 7 real points for all $0<x<1$, then π is ramified above 0 (projection onto the x-axis of a plane curve of the form $y^n+a_1(x)y^{n-1}+\ldots a_n(x)$ is a flat covering).

We leave to the reader to reformulate the Grothendieck-Hensel Lemma III.4.1.

Proposition R.14 Let $f: E \to X$ be a morphism of affine R-schemes, $e\in E(R)$ a point where f is smooth, $x = f(e)$. Then, there exists an affine R-scheme X', an etale morphism $\pi': X' \to X$, a point $x'\in X'(R)$ with $\pi(x') = x$ and a morphism $s: X' \to E$ such that $f\circ s = \pi$ and $s(x') = e$.

Examples of fields with an implicit function theorem.

The fraction field of a discrete henselian valuation ring by 7.4.

Fields with a complete norm: A topological field R has an implicit function theorem if and only if an analog of 7.4 holds for R. The uniqueness and continuity (with the terminology of 7.4) can be treated precisely as in the proof of 7.4. Existence: There exists a formal power series $S(X.)\in R[[X_1,\ldots,X_n]]$ without constant term such that

$$G(S(X.),X.) = 0 .$$

The norm of the coefficients of the power series satisfies the classical growth condition ($|s_{i.}| \leq \text{Const.} \sum_{t=1}^{n} \rho_t^{i_t}$) .

Chapter IV SOME FUNDAMENTAL THEOREMS

IV.1. Hilbert's Nullstellensatz

The first part of Hilbert's Nullstellensatz is

Theorem 1.1 Let k be a field and A a finite type k-algebra. If m is a maximal ideal in A, then A/m is of finite rank over k.

Proof Apply Lemma 1.2 below to the pair (m,1).
Lemma 1.2 below has a huge range of applications beyond 1.1.
If p is a prime ideal of a ring C, then k(p) denotes the fraction field of C/p.

Lemma 1.2 Let $f: A \to B$ be a morphism of finite type. Then, for all pairs (q,b), where q is a prime ideal of B and $b \in B-q$, there exists $a \in A-p\,(p = f^{-1}(q))$ such that

$$\forall \text{ primes } r \text{ of } A \text{ with } a \notin r \supseteq p$$
$$\exists \text{ prime } r' \text{ of } B \text{ with } b \notin r' \supseteq q$$

such that $f^{-1}(r') = r$ and $k(r) \to k(r')$ is finite algebraic.

Proof Consider the class of finite type morphisms for which the Lemma holds. This class is stable under composition and contains all surjective morphisms. Whence it suffices to prove the Lemma in case $B = A[X]$.

Replacing A by A/p we may assume that A is a domain and $q \cap A = 0$.

There are two independent cases to consider, $q = 0$, $q \neq 0$.

$q = 0$ Given $b \in A[X]$, $b \neq 0$. Write

$$b = \sum a_i X^{n-i} \quad , \; a_0 \neq 0$$

We are going to prove that a_0 will work for a in the Lemma.
The problem can be restated as follows: Given a morphism $t: A \to k$ from A to an algebraically closed field with $t(a_0) \neq 0$. Then there exists an extension t' of t to A[X] such that $t'(b) \neq 0$:
Extend t by mapping X to $x \in k$, where x is not a root of

$$\sum t(a_i) X^{n-i}$$

$q \neq 0$ Set $C = A[X]/q$ and let x, resp. c denote the residue class of X, resp. b. (K resp. L denotes the fraction field of A resp. C.)

$$\begin{array}{ccc} A & \subset & C \ni x, c \\ \cap & & \cap \\ K & \subset & L \ni c^{-1} \end{array}$$

C is generated by x as A-algebra and $K \to L$ is a finite extension.

Observe that if $y\in L$, then there exists $s\in A, s\neq o$ such that sy is integral over A: Let

$$\sum c_i y^{n-i} = o \quad \text{with} \quad c_i\in A,\ c_o\neq o.$$

Multiply that equation by c^{n-1} to see that $c_o y$ is integral over A.

An $a\in A, a\neq o$ such that ax and ac^{-1} are integral over A will work: Note first that C_{ac} is finite over A_a since it is generated by c^{-1} and x. By "lying over" any prime ideal r of A_a lifts to a prime ideal r' of C_{ac}.

Q.E.D.

The second part of Hilbert's Nullstellensatz is

Theorem 1.3 Let k be a field and A a finite type k-algebra. If A is reduced, then the zero ideal in A is intersection of maximal ideals.

Proof It is well known that the zero ideal in this case is intersection of prime ideals. Whence it suffices to prove that A is a Jacobsen ring, see below.

Definition 1.4 A ring R is called a Jacobsen ring if any prime ideal of R is intersection of maximal ideals.

Proposition 1.5 Let A denote a Jacobsen ring and $f: A \to B$ a morphism of finite type. Then

(i) B is a Jacobsen ring

(ii) If n is a maximal ideal of B, then $f^{-1}(n) = m$ is a maximal ideal of A and B/n is finite over A/m.

Proof (i) Let q be a prime ideal in B and $b\in B-q$. Let $a\in A-f^{-1}(q)$ be chosen according to Lemma 1.2. By assumption we can find a maximal ideal r of A such that $a\notin r$ and $r\supseteq f^{-1}(q)$. By 1.2 we can find $r'\supseteq q$ such that $b\notin r'$ and $k(r) \to k(r')$ is finite. This makes r' a maximal ideal.

(ii) Let q be a maximal ideal in B. Apply Lemma 1.2 to the pair (q,1) to get $a\in A-f^{-1}(q)$ with the property of the Lemma. Choose a maximal ideal $r\supseteq f^{-1}(q)$ with $a\notin r$. Then there exists a prime ideal $r'\supseteq q$ with $f^{-1}(r') = r$. This implies $r' = q$ and $r = f^{-1}(q)$. Q.E.D.

Proposition 1.6 Let A be a finite type k-algebra. If A only has finitely many maximal ideals, then A is of finite rank over k.

Proof We shall show that all prime ideals of A are maximal. Let q be a prime ideal. q is intersection of (finitely many) maximal ideals, $q = \bigcap m_i \supseteq \pi m_i$, whence $q \supseteq m_i$, some i. It follows that A is an Artin ring and whence of finite rank over k. Q.E.D.

Exercise 1.7 Let k denote an algebraically closed field. All k-algebras considered are of finite type over k. If C is a k-algebra, C(k) denotes $Hom_k(C,k)$. Similar if f: A → B is a k-morphism we define $f(k): B(k) \to A(k)$ by $f(k) = Hom_k(f,1_k)$. Thus C → C(k) defines a contravariant functor

$$\{k\text{-algebras}\} \Rightarrow \{sets\}$$

i) Show that if B is reduced and f and g are k-morphisms, then f(k) = g(k) implies f = g

ii) Let f: A → B be a k-morphism and i: A → C a surjective k-morphism. Show that if B is reduced, then f can be factored through i if and only if f(k) can be factored through i(k)

iii) Let C be a k-algebra. Show that C(k) is finite if and only if C is finite over k.

iv) Let C be a k-algebra. For an ideal $\mathfrak{a}$ of C let $V(\mathfrak{a})$ denote the image of $C/\mathfrak{a}(k) \to C(k)$. Show that the $V(\mathfrak{a})$'s satisfies the axioms for the closed sets in a topology, as $\mathfrak{a}$ runs through all ideals of C (the Zariski topology on C(k)).

v) Show that C → C(k) defines a covariant functor

$$\{k\text{-alg.}\} \Rightarrow \{top.\ spaces\}$$

which takes surjections into closed immersion

vi) Let C be a k-algebra. Show that $\mathfrak{a} \mapsto V(\mathfrak{a})$ establishes a one to one correspondance between ideals $\mathfrak{a}$ such that $A/\mathfrak{a}$ is reduced and closed subsets of C(k). Show that prime ideals correspond to the irreducible closed subsets.

vii) Establish a 1-1 correspondance between idempotents in C and subsets of C(k) which are open and closed.

viii) Recall that C(k) corresponds to maximal ideals in C. Thus we have a natural inclusion C(k) → Spec(C). Show that U → U∩C(k) defines a 1-1 correspondance between the open sets in Spec(C) and the open sets in C(k).

xi) For x∈C(k) let C_x denote the local ring at the corresponding maximal ideal. Let p be a prime ideal of C. Show that C_p is regular if and only if C_x is regular for x in a non empty open subset of V(p).

IV.2. Zariski's main theorem

Let $f: A \to B$ be a morphism, p a prime of A. $B \otimes_A k(p)$ is called the fiber of f at p. The primes in $B \otimes_A k(p)$ may be identified with the set of primes q in B such that $f^{-1}(p) = q$. q is said to be <u>isolated in its fiber</u> if q is as well maximal as minimal in $B \otimes_A k(p)$. - If f is of finite type, this means that $\{q\}$ is a connected component of $\mathrm{Spec}(B \otimes_A k(p))$

<u>Theorem 2.1</u> Let $A \subseteq B$ with A integrally closed in B such that there exists $t_1, \ldots, t_n \in B$ with B integral over $A[t_1, \ldots, t_n]$. If a prime ideal q of B is isolated over $p = q \cap A$, then there exists an $s \in A-p$ such that $A_s = B_s$.

<u>Proof</u> Hilbert's Nullstellensatz reduces immediately the number of variables to 1, see A below. So we have the situation

$$A \subseteq A[t] \subseteq B$$

with B integral over $A[t]$. Let f denote the conductor from B to $A[t]$, i.e.

$$f = \{v \in A[t] \mid vB \subseteq A[t]\}$$

Note that f is a B-ideal. The proof of the theorem in case $q \not\supseteq f$ and $q \supseteq f$ are completely different and independent, see B and C below.

<u>A Reduction to the case $n=1$ $(t=t_1)$</u>

The reduction is based on the following corollary to Hilbert's Nullstellensatz

<u>A.1</u> Let $R \subseteq S \subseteq T$ be such that T is integral over a finite type R-algebra. If a prime q_T of T is maximal in its fiber above $q_R = R \cap q_T$, then $q_S = S \cap q_T$ is maximal in its fiber over q_R.

Proof: We may assume R,S,T integral domain and $q = 0$. Next we may assume R is a field. This makes T a field. Let T' be a finite type sub-R-algebra of T such that T is integral over T'. T' must be a field since T is a field by the "lying over theorem".
By Nullstellensatz T' is (finite) algebraic over R whence T is algebraic over R this makes S a field. -

<u>A.2</u> Reduction to the case $n = 1$. Let C denote the integral closure of $A[t_1, \ldots, t_{n-1}]$ in B. $q_C = q \cap C$ is isolated over p: it is maximal above p by A.1, it is minimal by the inductive assumption, applied to $C \to B$. The rest is now straight forward.

B Case where $q \not\supseteq f$

We shall first prove the theorem in

B.1 Case where $B = A[t]$.
We may assume A local with maximal ideal p. Set $k = A/p$. In case $B \otimes_A k \simeq k[X]$ no primes above p are isolated in the fiber. So we may assume that there exists a polynomial $(n \geq o)$

$$a_{n+1}T^{n+1} + a_nT^n + \dots$$

with $a_i \in A$, at least one a_i is a unit in A and

$$a_{n+1}t^{n+1} + a_nt^n \dots = 0$$

We are going to prove that $t \in A$. Note, if $n = 0$ this is clear. In the general case note that $a_{n+1}t \in A$: multiply the above equation with a_{n+1}^n to see that $a_{n+1}t$ is integral over A. If a_{n+1} or $a_{n+1}t$ is a unit we get $t \in R$. If $a_{n+1}t$ and a_{n+1} belong to p we get an equation of lower degree

$$(a_n + a_{n+1}t)t^n + a_{n-1}t^{n-1} + \dots = 0$$

Conclusion by induction on n.

B.2 $q \not\supseteq f$, general case.
Set $r = q \cap A[t]$ and look at the injection $A[t]_r \longrightarrow B_r$. By assumption there exists $v \in A[t] - r$ such that $vB \subseteq A[t]$. That makes $A[t]_r \longrightarrow B_r$ an isomorphism, i.e. $B_r = B_q$ and $A[t]_r \xrightarrow{\sim} B_q$. This makes r isolated in the fiber above p, and whence there exists $s \in A - p$ by B.1 such that $A_s = A[t]_s$. B_s is integral over A_s on the one hand and A_s is integrally closed in B_s on the other. Whence $A_s = B_s$.

C Case where $q \supseteq f$

We are going to prove that no primes $q \supseteq f$ are isolated in their fibers. First an elementary calculation.

C.1 Let $C \subseteq C[t] \subseteq D$ where D is integral over $C[t]$. If the conductor f from D to $C[t]$ contains an element of the form $F(t)$ where F is a monic polynomial with coefficients in C, then $D = C[t]$, provided C is integrally closed in D.
Proof. Given $d \in D$. We are going to show that $d \in C[t]$. We have $F(t)d = G(t)$ for some $G \in C[T]$. Make an Euclidean division

$$G = QF + R \quad , \quad \text{dg } R < \text{dg } F$$

Put $s = d-Q(t)$. We are going to show $s \in C$. We know $F(t)s = R(t)$.

Assume for a moment that s is invertible in D (pass to D_s). Then

$$t \text{ is integral over } C[s^{-1}]$$
$$s \text{ is integral over } C[t]$$

This implies first that $s^m t$ is integral over C for some m. Multiply an integral dependence equation for s with respect to $C[t]$ by a high power of s^m to see that s is integral over C.

Returning to the original situation, we can find a monic $H \in C[T]$ such that $s^n H(s) = 0$ some n. Whence $s \in C$.

<u>C.2</u> With the notation of C.1, let n denote a minimal prime ideal of D/f. Then the image of t in D/n is transcendental over C/m, where $m = n \cap C$.

Proof. Replace C by its local ring at m. We may assume C local and m maximal: the conductor might not be preserved by localization but it is still contained in n as one easily checks. The proof is by contradiction. Assume t algebraic in D/n. Let $F \in C[T]$ be a monic polynomial with $F(t) \in n$. Since n is minimal with respect to the conductor f, B_n/f_n has only one prime ideal, whence some power of the image of $F(t)$ in B_n belongs to f_n, i.e. there exists $v \in B-n$ such that

$$v\,F(t)^d \in f \quad , \quad d \geq 1 \; .$$

Apply C.1 to $C[t,vB]$ and F^d to see that $v \in f$, which contradicts $f \subseteq n$.

Let us now return to our original situation $A \subseteq A[t] \subseteq B$ with B integral over $A[t]$. We are going to prove

<u>C.3</u> No prime ideal $q \supseteq f$ is isolated in its fiber.

Proof. Let n be a minimal prime ideal of B/f contained in q. Put $m = A \cap n$. The image of t in B/n is transcendental over A/m by C.2 whence replacing A by A/m and B by B/n it suffices to prove

<u>C.4</u> Let $A \subseteq A[T] \subseteq B$ where A and B are domains and B is integral over $A[T]$. Then no prime q of B is isolated in its fiber.

Proof. If $B = A[T]$, this is immediate. If A is normal, then $A[T]$ is normal and C.4 follows from the going up and going down theorems. In the general case introduce the normalizations of A and B in their fraction fields. Q.E.D.

Cor. 2.2 Let $A \rightarrow B$ be a local morphism essentially of finite type. Let k denote the residue field of A. If $k \otimes_A B$ is finite over k, then $A \rightarrow B$ is essentially finite.

Proof Choose a finite type subalgebra C of B and a prime ideal q of C above the maximal ideal of A, such that $B = C_q$. Let A' denote the normalization of A in C, and p' the retraction of q to A'. By 2.1, $A'_{p'} \rightarrow C_q$ is an isomorphism. Whence $A \rightarrow B$ is essentially integral, being essentially of finite type. It is consequently ess. finite.

Q.E.D.

Exercise 2.3 With the notation of 1.7, let $f: A \rightarrow B$ be a k-morphism, such that $f(k): B(k) \rightarrow A(k)$ has finite fibers. Show that $A \rightarrow B$ can be factored $A \rightarrow C \rightarrow B$ where $A \rightarrow C$ is finite and $C \rightarrow D$ is an open immersion (III.R.1).

IV.3. Chevalley's constructibility theorem

Definition 3.1 Let X be a topological space. A subset D of X is called locally closed if for each $x \in X$ there exists an open nbhd U such that $U \cap D$ is closed in U. It remains the same to say that D is the intersection of an open set of X and a closed set. - A subset C of X is called constructible if it is the finite union of locally closed sets.

Theorem 3.2 Let $A \rightarrow B$ be a morphism of finite type with A noetherian. Then the image of $\mathrm{Spec}(B) \rightarrow \mathrm{Spec}(A)$ is a constructible set.
We shall first prove an important characterization of constructible set in a noetherian topological space (Bourbaki, Alg. Comm. II, §4).

Notation 3.3 Let C denote an irreducible subspace of the topological space X, P any subset of X. We say that almost all points of C are in P, if $P \cap C$ contains a non empty open subset of C.

Lemma 3.4 Let D denote a subset of the noetherian topological space X. D is constructible if and only if, whenever F is an irreducible closed subset of X, then either almost all points of F are in D or almost all points of F are in X-D.

Proof Let F be a closed irreducible subset of X. Consider the set of subsets D' of X such that almost all points of F are in X-D' or in D'.

That set of subsets is stable under finite unions and finite intersections and contains all closed and open sets, whence contains all constructible sets.

Next, suppose D has the second property above. Let us prove that D is constructible in X under the additional assumption that for all closed proper subsets F of X, $F \cap D$ is constructible in F (or X, that does not matter).

Let us examine the various possibilities

1. X irreducible

1.1 Almost all points of X are in D. Let U be a non empty open subset of X, $U \neq X$ and $U \subseteq D$. Then $D = U \cup ((X-U) \cap D)$ is constructible.

1.2 Almost all points of X are in X-D. Then $\bar{D} \neq X$ and whence D constructible in $\bar{D}$ and thus in X.

2. X is not irreducible.

   Write $X = F \cup E$ where F and E are closed sets distinct from X.

   $D = (D \cap E) \cup (D \cap F)$ is constructible.

We can now finish the proof by Noetherian induction: suppose D is not constructible. Then there exists a smallest closed set X_0 of X such that $D_0 = X_0 \cap D$ is not constructible in X_0. For all proper closed subsets F_0 of X_0 we have that $D_0 \cap F_0$ is constructible. Q.E.D.

Proof of Chevalley's theorem

Let Y = Spec(A) and X = Spec(B). I denotes the image of $f: X \rightarrow Y$.

We are going to prove the following

1^0 If an irreducible, closed subset D of Y is of the form $\overline{f(C)}$ where C is an irreducible, closed subset of X, then almost all points of D are in I

2^0 If an irreducible closed subset D of Y is not of the form $\overline{f(C)}$, C irreducible, closed subset of X, then almost all points of D are in Y-I

$\underline{1^0}$ Let q denote the generic point for C i.e. $C = \overline{\{q\}}$ and p the generic point for D. Apply Lemma 1.2 to the pair (q,1) to get that almost all points of D are in the image of C and whence in I.

$\underline{2^0}$ We have $I \cap D = f(f^{-1}(D))$. Write $f^{-1}(D) = \cup C_i$ where the C_i's are closed irreducible subsets of X. We have $I \cap D = \cup f(C_i)$ and whence $I \cap D \subseteq \cup \overline{f(C_i)} \subseteq D$. We have $\cup \overline{f(C_i)} \neq D$, otherwise $D = \overline{f(C_i)}$ for some i contradicting the assumption on D. Thus the complement of $\bigcup f(C_i)$ in D does not meet I. Q.E.D.

Proposition 3.5 Let D be a dense constructible set of the noetherian topological space X. Then D contains a dense open set of X.

Proof Let $X_1,\ldots,X_r$ denote the irreducible components of X. Let O_i denote the complement to $\bigcup_{j,j\neq i} X_j$ in X. O_i is a non empty open set of X contained in X_i. - Put $D_i = D\cap X_i$. D_i is dense in X_i since D is dense in X and X_i is an irreducible component of X. D_i is constructible and whence (3.4) almost all points of X_i are contained in D_i. Whence there exists an open set U_i of X, such that $\emptyset \neq U_i \cap D$. $W_i = O_i \cap U_i$ is a non empty open set of X contained in X_i and D. $\bigcup W_i$ is open and dense.

Q.E.D.

Lemma 3.6 Let f: A → B be a flat morphism. Let q be a prime ideal in B and p' a prime ideal in A contained in $p = f^{-1}(q)$. Then there exists a prime ideal $q' \subseteq q$ such that $f^{-1}(q') = p'$.

Proof Replace A by A_p and B by B_q. This makes A → B a flat local morphism, and whence faithfully flat by Bourbaki, Alg. Comm. I, §3, n°5, Prop. 9. Conclusion by loc. cit. II, §2, n°5, Cor. 4 de la prop. 11.

Q.E.D.

Proposition 3.7 Let f: A → B be a flat morphism of finite type between noetherian rings. Then

$$\mathrm{Spec}(B) \longrightarrow \mathrm{Spec}(A)$$

is an open map.

Proof It suffices to prove that the image I is open. Let C denote the complement to I. We are going to prove $C = \overline{C}$. I and whence C is constructible by 3.2. Let $D_1,\ldots,D_m$ be the irreducible components of $\overline{C}$, and d_i the generic point of D_i. $d_i \in C$ by 3.5 and any prime of Spec(A) containing d_i is in C by 3.5, whence $D_i \subseteq C$. Q.E.D.

Remark 3.8 Let R be a noetherian ring. A subset S of Spec(A) is said to be stable under specialization if $\overline{\{s_1\}} \supseteq \overline{\{s_2\}}$ and $s_1 \in S \Rightarrow s_2 \in S$. The proof of 3.7 show that a constructible subset of Spec(A), stable under specialization, is closed.

Remark 3.9 The proof of 3.7 shows more generally that if f: A → B is a finite type morphism between noetherian ring with a "going down theorem" i.e. satisfies Lemma 3.6, then Spec(B) $\to$ Spec(A) is open.

Exercise 3.10 With the notion of 1.7. Let $f: A \to B$ be a k-morphism. Show that the image of $f(k): B(k) \to A(k)$ is a constructible set. Show that if f is flat then f(k) is an open morphism.

Exercise 3.11 Let $f: A \to B$ be finite and flat. Show that

$$x \mapsto \# f(k)^{-1}(x)$$

is a lower semi-continuous function on A(k). Hint: use 3.10 and the theory of henselization as it is applied in III. R.12.

IV C The analytic case

C. 1 Noether normalization lemma

Let k be a field and A a finite type k-algebra which is a domain. Then there exists algebraically independent elements $t_1,\ldots,t_n$ of A such that A is finite over $k[t_1,\ldots,t_n]$.

Proof Let A be generated by $x_1,x_2,\ldots,x_r$. The theorem is proved by induction on the number of generators r. It is obviously true for r = 1. Assume the theorem is true for all domains generated by r-1 elements.

We may assume that we have a relation of the form

$$\sum_\nu a_\nu x_1^{\nu(1)} x_2^{\nu(2)} \ldots x_r^{\nu(r)} = 0$$

with $a_\nu \neq 0$ for some $\nu \neq (0,\ldots,0)$. Choose an integer g such that

$$g > \nu(j), j = 1,\ldots,r, \text{ all } \nu \text{ with } a_\nu \neq 0$$

and set

$$z_i = x_i - x_1^{g^{i-1}}$$

we have

$$\sum_\nu a_\nu x_1^{\nu(1)}(z_2+x_1^g)^{\nu(2)}(z_3+x_1^{g^2})^{\nu(3)}\ldots = 0$$

The term of highest degree in x_1 is of the form

$$a_\nu x_1^{\nu(1)+g\nu(2)+g^2\nu(3)+\ldots}$$

Note that the exponents to x_1 are different for different ν's since they have different g-adique expansions. Whence x_1 is integral over $k[z_2,\ldots,z_r]$ Q.E.D.

From now on we shall work with affine schemes over $\mathbb{C}$, see III. R.

An affine scheme whose coordinate ring is an integral domain is called an <u>affine variety</u>. - By the Noether normalization Lemma we have

<u>C.2</u> An affine variety of dimension n admits a finite surjective morphism onto $\mathbb{A}^n$.

<u>Cor. C.3</u> If X is an affine $\mathbb{C}$-scheme such that $X(\mathbb{C})$ is compact in the analytic topology, then X is finite over $\mathbb{C}$.

<u>Proof</u> It suffices to show that all irreducible components of X are finite over $\mathbb{C}$. So we may assume that X is an affine variety. We can find a finite surjective morphism $X \to \mathbb{A}^n$. $X(\mathbb{C})$ compact makes $\mathbb{C}^n$ compact, whence $n = o$.

<u>Theorem C.4</u> Let X be an affine variety over $\mathbb{C}$. A non empty Zariski open subset of $X(\mathbb{C})$ is dense in $X(\mathbb{C})$ in the analytic topology.

We shall first need two Lemmas.

<u>Lemma C.5</u> Let $f: X \to Y$ be a finite morphism. Then $X(\mathbb{C}) \to Y(\mathbb{C})$ is a proper map in the analytic topology.

Since the spaces involved are locally compact this means that the inverse image of a compact subset is compact.

<u>Proof</u> Imbed Y into $\mathbb{A}^m$ to see that we may assume $Y = \mathbb{A}^m$. Imbed X into $\mathbb{A}^n$ and let $f_1,\ldots,f_m$ be polynomials in n-variables such that the restriction of

$$x. \longmapsto (f_1(x.),\ldots,f_m(x.))$$

to $X(\mathbb{C})$ is $f(\mathbb{C})$.

The coordinate functions $x_1,\ldots,x_n$ satisfies equations of the form

$$x_j^q + P_{1.j}(f.(x.))x_j^{q-1} + \ldots + P_{q.j}(f.(x.)) = o$$

for all $x. \in X(\mathbb{C})$ where $P_{r.s} \in \mathbb{C}[X_1,\ldots,X_m]$. Whence it is sufficient to prove: let

$$T^q + a_1 T^{q-1} + \ldots a_q$$

be a polynomial with coefficients in $\mathbb{C}$. If $|a_i| \leq D$ then there exists a bound for the size of the roots of the above polynomial, depending only on D. This is easy and left to the reader. Q.E.D.

<u>Lemma C.6</u> Let X be a variety and $f: X \to \mathbb{A}^n$ a finite surjective morphism. Let $x \in X(\mathbb{C})$ and $y_1,\ldots,y_p,\ldots$ a sequence of points in $\mathbb{C}^n$ which converges towards $f(x)$. Then there exists a sequence of points $x_1,\ldots,x_p,\ldots$ in $X(\mathbb{C})$ which converges towards x and such that

$f(x_i) = y_i$ for all i.

Proof Let us first treat the case where $X = \mathbb{F}^n = \{(t,a.)\in \mathbb{A}^1 \times \mathbb{A}^n \mid t^n + a_1 t^{n-1} + a_2 t^{n-2} + \ldots = 0\}$ and $f: \mathbb{F}^n \to \mathbb{A}^n$ the canonical projection. By translation we can assume x of the form $(0,a.)$. Put $y_i = a.^{(i)}$. $(a_n^{(i)})_n$ converges towards zero and $a_n^{(i)}$ is the product of the roots of the equation

$$t^n + a_1^{(i)} t^{n-1} + a_2^{(i)} t^{n-2} + \ldots = 0 ,$$

thus we can find a root $t^{(i)}$ such that $t^{(i)} \to 0$ as i goes to infinity.

In the general case let $x_1, x_2, \ldots, x_r$ be the points in the fiber of $X(\mathbb{C}) \to \mathbb{A}^n(\mathbb{C})$ at $f(x)$. Let A denote the affine coordinate ring for X. Choose $a \in A$ such that $a(x) = 0$ and $a(x_i) \neq 0$ $i \geq 2$. Let

$$T^m + h_1 T^{m-1} + \ldots h_m$$

be minimal polynomial of a with respect to $k[X_1, \ldots, X_n]$. This polynomial is irreducible since $k[X_1, \ldots, X_n]$ is factorial. Whence we have $h: \mathbb{A}^n \to \mathbb{A}^m$ and a factorization of f

$$X \xrightarrow{g} h^* \mathbb{F}^m \longrightarrow \mathbb{A}^n$$

with x being the only point in its fiber with respect to g.

Lift $y_1, \ldots, y_p, \ldots$ to a sequence $x_1', \ldots, x_p'$ in $h^* \mathbb{F}^m$ which converges towards $g(x)$ which we may according to the preceeding consideration. Let $x_1, \ldots, x_p, \ldots$ be an arbitrary lifting of $x_1', \ldots, x_p', \ldots$. - $x_1, \ldots, x_p, \ldots$ converges towards x: the only possible cluster point for $x_1, \ldots, x_p, \ldots$ is x since x is the only point in its g-fiber and $g(x_1), \ldots, g(x_p), \ldots$ converges towards $g(x)$. Conclusion by C.5 Q.E.D.

Proof of C.4 Case $X = \mathbb{A}^n$.

Let O be a non empty Zariski open subset of $\mathbb{C}^n$ and y a point of the complement V of O in $\mathbb{C}^n$. We can find a non constant polynomial P in n variables which is zero on V. Let z be a point such that $P(z) \neq 0$.

$$P(Ty + (1-T)z)$$

is non constant. Whence we can find a sequence $y_1, \ldots, y_p, \ldots$ in O which converges towards y.

General case Let U be a non empty Zariski open subset of $X(\mathbb{C})$ and x a point of the complement F of U in $X(\mathbb{C})$. Choose a finite surjective morphism

$$f: X \to \mathbb{A}^n$$

$f(F)$ is Zariski closed in $\mathbb{C}^n$ and distinct from $\mathbb{C}^n$. Choose a sequence of points in the complement of $f(F)$ in $\mathbb{C}^n$ which converges towards $f(x)$. Lift the sequence to a sequence on $X(\mathbb{C})$ which converges towards x which we may by C.6. Q.E.D.

BIBLIOGRAPHY

ARTIN, M.: Grothendieck Topologies. Mimeographed Seminar Notes. Harvard 1962.

ARTIN, M.: Algebraic approximation of structures over complete local rings, Publ. Math. I.H.É.S. n°36 (1969), 23-58

AUSLANDER, M. and BUCHSBAUM, D.A.: On ramification thoory in noetherian rings. Amer. J. Math. 81 (1959), 749-765

AUSLANDER, M. and GOLDMANN, O.: The Brauer group of a commutative ring. Trans. Amer. Math. Soc. 97 (1960), 367-409

AZUMAYA, G.: On maximally central algebras. Nagoya Math. J. 2 (1951) 119-150

BOREL, A.: Linear Algebraic Groups. Benjamin, New York 1969

BOURBAKI, N.: Algèbre I. Hermann, Paris 1970

BOURBAKI, N.: Algèbre Commutative. Chap. 1,2. Hermann, Paris 1961. Chap. 5. Hermann, Paris 1964

CHEVALLEY, C.: Fondements de la géométrie algébrique. Secrétariat Mathématique, Paris 5e 1958

COHEN, I.S.: On the structure and ideal theory of complete local rings. Trans. Amer. Math. Soc. 59 (1946), 54-106

DEMAZURE, M. et GABRIEL, P.: Groupes Algébriques. Tome I. Masson & Cie., North-Holland, Paris-Amsterdam 1970

EVANS, E.G. Jr.: A generalization of Zariski's main theorem. To appear

GREENBERG, M.: Rational points in henselian discrete valuation rings. Publ. Math. I.H.É.S. n°31 (1966), 59-64

GROTHENDIECK, A.: Revêtements Etales et Groupe Fondamental. Lecture Notes in Mathematics 224, Springer 1971

GROTHENDIECK, A. et DIEUDONNÉ, J.: Éléments de Géométrie Algébrique. Publ.Math. I.H.É.S. n°4,8,11,17,20,24,28,32 (1960 ff)

GROTHENDIECK, A.: Le groupe de Brauer I. Séminaire Bourbaki 1964/65, n°290

GROTHENDIECK, A.: Schemas en Groupes I-II. Lecture Notes in Mathematics 151, 152, Springer 1970

HOCHSCHILD, G.: On the cohomology groups of an associative algebra. Ann. of Math. 46 (1945), 58-67

LANG, S.: Introduction to Algebraic Geometry. Interscience Publ., New York 1958

LICHTENBAUM, S. and SCHLESSINGER, M.: The cotangent complex of a morphism. Trans. Amer. Math. Soc. 128 (1967), 41-70

MATSUMURA, H.: Commutative Algebra. Benjamin, New York 1970

MUMFORD, D.: Introduction to algebraic geometry. Harvard Lecture Notes

NAGATA, M.: On the theory of Henselian rings. Nagoya Math. J. 5 (1953), 45-57

NAGATA, M.: On the theory of Henselian rings, II. Nagoya Math. J. 7 (1954), 1-19

NAGATA, M.: On the theory of Henselian rings, III. Mem. Coll. Sci. Univ. Kyoto 32 (1959-1960), 93-101

ONO, T.: Arithmetic of algebraic tori. Annals of Math. 74 (1961), 101-139

PESKINE, C.: Une généralization du "Main Theorem" de Zariski. Bull. Sc. Math. 2e series 90 (1966), 119-127

QUILLEN, D.: On the (Co-)Homology of Commutative Rings. Proc. of Symp. in pure Math. XVII, 65-87, Amer. Math. Soc. 1970

RAYNAUD, M.: Anneaux Henséliens. Lecture Notes in Mathematics 169, Springer 1970

ROSENBERG, A. and ZELINSKY, D.: Automorphisms of separable algebras. Pac. J. Math. 11 (1961), 1109-1117

SCHLESSINGER, M.: Functors of Artin rings. Trans. Amer. Math. Soc. 130 (1968), 208-222

SERRE, J.-P.: Géométrie algébriques et géométrie analytiques. Ann. Inst. Fourier 6 (1955-56), 1-42

WEIL , A.: Foundations of Algebraic Geometry. Amer. Math. Soc., New York 1946

ZARISKI, O.: Foundation of a general theory of birational correspondances. Trans. Amer. Math. Soc. 53 (1943), 490-542

ZARISKI, O.: The concept of a simple point of an abstract algebraic variety. Trans. Amer. Math. Soc. 62 (1947), 1-52

ZARISKI, O. and SAMUEL,P.: Commutative Algebra. 2 vols., van Nostrand, Princeton 1958, 1960

Séminaire H. CARTAN et C. CHEVALLEY, 8e année: 1955/56, Secrétariat mathématique, Paris 5e

INDEX OF NOTATIONS

INDEX OF SYMBOLS

Vol. 178: Th. Bröcker und T. tom Dieck, Kobordismentheorie. XVI, 191 Seiten. 1970. DM 18,–

Vol. 179: Seminaire Bourbaki – vol. 1968/69. Exposés 347-363. IV. 295 pages. 1971. DM 22,–

Vol. 180: Séminaire Bourbaki – vol. 1969/70. Exposés 364-381. IV, 310 pages. 1971. DM 22,–

Vol. 181: F. DeMeyer and E. Ingraham, Separable Algebras over Commutative Rings. V, 157 pages. 1971. DM 16.–

Vol. 182: L. D. Baumert. Cyclic Difference Sets. VI, 166 pages. 1971. DM 16,–

Vol. 183: Analytic Theory of Differential Equations. Edited by P. F. Hsieh and A. W. J. Stoddart. VI, 225 pages. 1971. DM 20,–

Vol. 184: Symposium on Several Complex Variables, Park City, Utah, 1970. Edited by R. M. Brooks. V, 234 pages. 1971. DM 20,–

Vol. 185: Several Complex Variables II, Maryland 1970. Edited by J. Horváth. III, 287 pages. 1971. DM 24,–

Vol. 186: Recent Trends in Graph Theory. Edited by M. Capobianco/ J. B. Frechen/M. Krolik. VI, 219 pages. 1971. DM 18.–

Vol. 187: H. S. Shapiro, Topics in Approximation Theory. VIII, 275 pages. 1971. DM 22,–

Vol. 188: Symposium on Semantics of Algorithmic Languages. Edited by E. Engeler. VI, 372 pages. 1971. DM 26,–

Vol. 189: A. Weil, Dirichlet Series and Automorphic Forms. V. 164 pages. 1971. DM 16,–

Vol. 190: Martingales. A Report on a Meeting at Oberwolfach, May 17-23, 1970. Edited by H. Dinges. V, 75 pages. 1971. DM 16,–

Vol. 191: Séminaire de Probabilités V. Edited by P. A. Meyer. IV, 372 pages. 1971. DM 26,–

Vol. 192: Proceedings of Liverpool Singularities – Symposium I. Edited by C. T. C. Wall. V, 319 pages. 1971. DM 24,–

Vol. 193: Symposium on the Theory of Numerical Analysis. Edited by J. Ll. Morris. VI, 152 pages. 1971. DM 16,–

Vol. 194: M. Berger, P. Gauduchon et E. Mazet. Le Spectre d'une Variété Riemannienne. VII, 251 pages. 1971. DM 22,–

Vol. 195: Reports of the Midwest Category Seminar V. Edited by J.W. Gray and S. Mac Lane. III, 255 pages. 1971. DM 22,–

Vol. 196: H-spaces – Neuchâtel (Suisse)- Août 1970. Edited by F. Sigrist, V, 156 pages. 1971. DM 16,–

Vol. 197: Manifolds – Amsterdam 1970. Edited by N. H. Kuiper. V, 231 pages. 1971. DM 20,–

Vol. 198: M. Hervé, Analytic and Plurisubharmonic Functions in Finite and Infinite Dimensional Spaces. VI, 90 pages. 1971. DM 16.–

Vol. 199: Ch. J. Mozzochi, On the Pointwise Convergence of Fourier Series. VII, 87 pages. 1971. DM 16,–

Vol. 200: U. Neri, Singular Integrals. VII, 272 pages. 1971. DM 22,–

Vol. 201: J. H. van Lint, Coding Theory. VII, 136 pages. 1971. DM 16,–

Vol. 202: J. Benedetto, Harmonic Analysis on Totally Disconnected Sets. VIII, 261 pages. 1971. DM 22,–

Vol. 203: D. Knutson, Algebraic Spaces. VI, 261 pages. 1971. DM 22,–

Vol. 204: A. Zygmund, Intégrales Singulières. IV, 53 pages. 1971. DM 16,–

Vol. 205: Séminaire Pierre Lelong (Analyse) Année 1970. VI, 243 pages. 1971. DM 20,–

Vol. 206: Symposium on Differential Equations and Dynamical Systems. Edited by D. Chillingworth. XI, 173 pages. 1971. DM 16,–

Vol. 207: L. Bernstein, The Jacobi-Perron Algorithm – Its Theory and Application. IV, 161 pages. 1971. DM 16,–

Vol. 208: A. Grothendieck and J. P. Murre, The Tame Fundamental Group of a Formal Neighbourhood of a Divisor with Normal Crossings on a Scheme. VIII, 133 pages. 1971. DM 16,–

Vol. 209: Proceedings of Liverpool Singularities Symposium II. Edited by C. T. C. Wall. V, 280 pages. 1971. DM 22,–

Vol. 210: M. Eichler, Projective Varieties and Modular Forms. III, 118 pages. 1971. DM 16,–

Vol. 211: Théorie des Matroïdes. Edité par C. P. Bruter. III, 108 pages. 1971. DM 16,–

Vol. 212: B. Scarpellini, Proof Theory and Intuitionistic Systems. VII, 291 pages. 1971. DM 24,–

Vol. 213: H. Hogbe-Nlend, Théorie des Bornologies et Applications. V, 168 pages. 1971. DM 18,–

Vol. 214: M. Smorodinsky, Ergodic Theory, Entropy. V, 64 pages. 1971. DM 16,–

Vol. 215: P. Antonelli, D. Burghelea and P. J. Kahn, The Concordance-Homotopy Groups of Geometric Automorphism Groups. X, 140 pages. 1971. DM 16,–

Vol. 216: H. Maaß, Siegel's Modular Forms and Dirichlet Series. VII, 328 pages. 1971. DM 20,–

Vol. 217: T. J. Jech, Lectures in Set Theory with Particular Emphasis on the Method of Forcing. V, 137 pages. 1971. DM 16,–

Vol. 218: C. P. Schnorr, Zufälligkeit und Wahrscheinlichkeit. IV, 212 Seiten 1971. DM 20,–

Vol. 219: N. L. Alling and N. Greenleaf, Foundations of the Theory of Klein Surfaces. IX, 117 pages. 1971. DM 16,–

Vol. 220: W. A. Coppel, Disconjugacy. V, 148 pages. 1971. DM 16,–

Vol. 221: P. Gabriel und F. Ulmer, Lokal präsentierbare Kategorien. V, 200 Seiten. 1971. DM 18,–

Vol. 222: C. Meghea, Compactification des Espaces Harmoniques. III, 108 pages. 1971. DM 16,–

Vol. 223: U. Felgner, Models of ZF-Set Theory. VI, 173 pages. 1971. DM 16,–

Vol. 224: Revètements Etales et Groupe Fondamental. (SGA 1). Dirigé par A. Grothendieck XXII, 447 pages. 1971. DM 30,–

Vol. 225: Théorie des Intersections et Théorème de Riemann-Roch. (SGA 6). Dirigé par P. Berthelot, A. Grothendieck et L. Illusie. XII, 700 pages. 1971. DM 40,–

Vol. 226: Seminar on Potential Theory, II. Edited by H. Bauer. IV, 170 pages. 1971. DM 18,–

Vol. 227: H. L. Montgomery, Topics in Multiplicative Number Theory. IX, 178 pages. 1971. DM 18,–

Vol. 228: Conference on Applications of Numerical Analysis. Edited by J. Ll. Morris. X, 358 pages. 1971. DM 26,–

Vol. 229: J. Väisälä, Lectures on n-Dimensional Quasiconformal Mappings. XIV, 144 pages. 1971. DM 16,–

Vol. 230: L. Waelbroeck, Topological Vector Spaces and Algebras. VII, 158 pages. 1971. DM 16,–

Vol. 231: H. Reiter, L^1-Algebras and Segal Algebras. XI, 113 pages. 1971. DM 16,–

Vol. 232: T. H. Ganelius, Tauberian Remainder Theorems. VI, 75 pages. 1971. DM 16,–

Vol. 233: C. P. Tsokos and W. J. Padgett. Random Integral Equations with Applications to Stochastic Systems. VII, 174 pages. 1971. DM 18,–

Vol. 234: A. Andreotti and W. Stoll. Analytic and Algebraic Dependence of Meromorphic Functions. III, 390 pages. 1971. DM 26,–

Vol. 235: Global Differentiable Dynamics. Edited by O. Hájek, A. J. Lohwater, and R. McCann. X, 140 pages. 1971. DM 16,–

Vol. 236: M. Barr, P. A. Grillet, and D. H. van Osdol. Exact Categories and Categories of Sheaves. VII, 239 pages. 1971. DM 20,–

Vol. 237: B. Stenström. Rings and Modules of Quotients. VII, 136 pages. 1971. DM 16,–

Vol. 238: Der kanonische Modul eines Cohen-Macaulay-Rings. Herausgegeben von Jürgen Herzog und Ernst Kunz. VI, 103 Seiten. 1971. DM 16,–

Vol. 239: L. Illusie, Complexe Cotangent et Déformations I. XV, 355 pages. 1971. DM 26,–

Vol. 240: A. Kerber, Representations of Permutation Groups I. VII, 192 pages. 1971. DM 18,–

Vol. 241: S. Kaneyuki, Homogeneous Bounded Domains and Siegel Domains. V, 89 pages. 1971. DM 16,–

Vol. 242: R. R. Coifman et G. Weiss, Analyse Harmonique Non-Commutative sur Certains Espaces. V, 160 pages. 1971. DM 16,–

Vol. 243: Japan-United States Seminar on Ordinary Differential and Functional Equations. Edited by M. Urabe. VIII, 332 pages. 1971. DM 26,–

Vol. 244: Séminaire Bourbaki – vol. 1970/71. Exposés 382–399. IV, 356 pages. 1971. DM 26,–

Vol. 245: D. E. Cohen, Groups of Cohomological Dimension One. V, 99 pages. 1972. DM 16,–